MINERAL TREASURES OF THE OZARKS

4880 Lower Valley Road • Atglen, PA 19310

Other Schiffer Books by the Author:

Cenozoic Fossils 1: Paleogene
978-0-7643-3424-5

Cenozoic Fossils II: The Neogene
978-0-7643-3580-8

Jewels of the Early Earth: Minerals and Fossils of the Precambrian
978-0-7643-3880-9

Mesozoic Fossils II: The Cretaceous Period
978-0-7643-3259-3

Mesozoic Fossils: Triassic and Jurassic
978-0-7643-3163-3

Meteorites
978-0-7643-3728-4

More Paleozoic Fossils
978-0-7643-4030-7

Paleozoic Fossil Plants
978-0-7643-4327-8

World's Oldest Fossils
978-0-7643-2697-4

Other Schiffer Books on Related Subjects:

Library of Congress Control Number: 2014944668

Designed by Molly Shields
Type set in Interstate-BoldCondensed/ITC Souvenir

ISBN: 978-0-7643-4715-3
Printed in China

Published by Schiffer Publishing, Ltd.
4880 Lower Valley Road
Atglen, PA 19310
Phone: (610) 593-1777; Fax: (610) 593-2002
E-mail: Info@schifferbooks.com

CONTENTS

INTRODUCTION

Mississippi Valley Type (MVT) Minerals and Mineral Deposits

This is not the usual picture book of minerals, as only a specific group of minerals that formed under rather specific conditions are considered. Minerals presented here come from what geologists call stratiform mineral deposits, and MVT minerals come from a specific type of stratiform mineral deposit that occurs extensively in the watershed of the Mississippi River, hence the name: Mississippi Valley Type (MVT) Minerals. Some of these deposits have produced valuable materials for over three centuries; this is especially the case with the mining of lead. Other collectable MVT minerals are those of zinc and copper, and more recently barite, the latter which, until about 100 years ago, had little value as a mineral commodity.

Presented here are minerals primarily associated with the Ozark Uplift of Missouri and Arkansas, thus there is a bias in this work toward minerals from stratiform deposits of the U.S. Midwest. This presentation also is different from the more conventional method of presenting minerals, which often has as its common denominator, something to do with chemical composition. My previous Schiffer books have dealt with fossils and I have been questioned for not sticking to this topic. My rebuttal to this is that there is a parallelism between fossils and MVT minerals. Many of the minerals discussed here follow a stratigraphic continuity similar to that associated with and characteristic of fossils. Also, a more obvious similarity with fossils is that MVT minerals are associated with and found in sedimentary rocks.

The more general, and less specific name for mineral occurrences discussed here—that is, mineral deposits that occur in sedimentary rocks—is epithermal deposits. Epithermal

EON	ERA	PERIOD	AGE MY.
PHANEROZOIC	CENOZOIC	TERTIARY	0 67
	MESOZOIC	CRETACEOUS	
		JURASSIC	
		TRIASSIC	235
	PALEOZOIC	PERMIAN	280
		PENNSYLVANIAN	
		MISSISSIPPIAN	300
		DEVONIAN	
		SILURIAN	350 320
		ORDOVICIAN	
		OZARKIAN-CANADIAN	440
		CAMBRIAN	500
PRECAMBRIAN ↓			542

Geologic time scale (with Paleozoic Era). MVT (Mississippi Valley Type) mineral deposits are predominantly associated with rocks formed during the Paleozoic Era. Different types of minerals can be associated with specific parts of geologic time somewhat in the same manner that fossils are indicative of specific portions of geologic time. This is one of the peculiarities of MVT mineral deposits. MVT mineral occurrences of the U.S. Midwest occur predominantly in Cambrian, Ordovician, and Mississippian rocks of the Mississippi Valley drainage basin.

Thick, horizontal dolostone beds of Cambrian age–Missouri Ozarks. This is the usual type of rock that hosts MVT minerals.

Thick sequences of carbonate rock (limestone and dolomite or dolostone) are a prerequisite for MVT mineral deposits. Usually, these rocks were deposited during the Paleozoic Era of geologic time. This thick sequence of limestone-like rock known as dolostone, has conduits dissolved in it that now carry groundwater to the surface, forming one of the largest springs in North America and known as Big Spring. Located in the "Ozark National Scenic Riverways" of Missouri, Big Spring is just one of many large springs that feed into the Current River, a typical Ozark stream. The age of the rock through which the spring flows is Cambrian; it was deposited during the Cambrian period of geologic time.

Horizontal beds of Cambrian age sandstone that underlie dolostone and limestone sequences that might contain MVT mineral deposits.

Geological environment in which MVT minerals may be found. Road construction in 2010 cut into massive beds of calcite incorporated into a thick bed of Cambrian dolostone. The white calcite can be seen as veins and masses in the tan dolostone. No mineralization was evident, but similar masses of calcite over the Ozarks (and elsewhere) can locally be charged with minerals of lead (galena), zinc (sphalerite), and copper (chalcopyrite) in enough quantity to make an ore body worth mining. Some of the largest occurrences of these three minerals are associated with such deposits. This type of deposit is named after the Mississippi Valley watershed and is known as Mississippi Valley Type deposits (MVT deposits).

Late Paleozoic (Mississippian Period) slabby limestone exposed along a river (Salt River) in northeastern Missouri. Original investigations for mineral deposits in the U.S. Midwest were done by floating rivers like this in small boats to access and examine the rocks exposed along them—mapping geology and looking for the presence of mineralization. Often this was done before statehood was established so that mineral lands could be set aside for mineral production instead of being offered for homesteading. One of the early MVT mineral areas in the upper Midwest was discovered in this way near the Mississippi River. This region, later known as the northern Tri-State mining region, began producing lead minerals (galena) just after the Louisiana Purchase. The town of Galena, Illinois, is named after this mineralized area.

Close up of calcite masses in the previous road cutting.

Quartz veins like these are indicative of hydrothermal mineral deposits. Hydrothermal mineral deposits are those whose mineralization originated from vapors given off by masses of molten rock. Molten rock injected into the earth's crust in what are known as igneous intrusions. MVT mineral deposits have no relationship with igneous rock or quartz veins.

mineral deposits form relatively close to the earth's surface and contrast with hyperthermal ones, which form at depth, usually from vapors or fluids associated with igneous phenomena—igneous phenomena which took place deep beneath the earth. Originally, mineral deposits of the Mississippi Valley area, as well as similar deposits occurring elsewhere, were assumed to have formed from deep-seated igneous activity: deep-seated igneous activity, which gave off water-rich vapors containing valuable metals associated with it. Such a hydrothermal (water and heat) origin of mineral deposits is the norm used worldwide to explain deposits of heavy metals like silver and gold, which are generally found in quartz veins. This also seemed to be the best explanation for those mineral deposits of the Midwest, although igneous or volcanic activity was not found to be associated with them.

MVT Mineral Deposits and Early Mining in North America

Mining of near-surface minerals (especially galena) in the Louisiana Territory of France in the 18th century established a portion of the Mississippi Valley Region (especially the northeastern portion of the Ozark Uplift) as a significant region for mining and mineral production. By the 19th century, MVT-type minerals had been found over a large portion of the Ozarks—areas which included, besides the Washington County lead/barite region, the "Old lead belt" near Bonneterre and Park Hills (Flat River) Missouri, the Tri-State area of Missouri, Oklahoma, and Kansas and the fluorspar region of southern Illinois—all designated to become world-class mining areas endowed with rich mineralization.

Another group of quartz veins. MVT minerals and other epithermal mineral deposits are never associated with quartz veins. Quartz veins are characteristic of hydrothermal mineral deposits.

Mine au Breton, Louisiana Territory, late 1700s. One of the first mining areas in North America, established c. 1730.

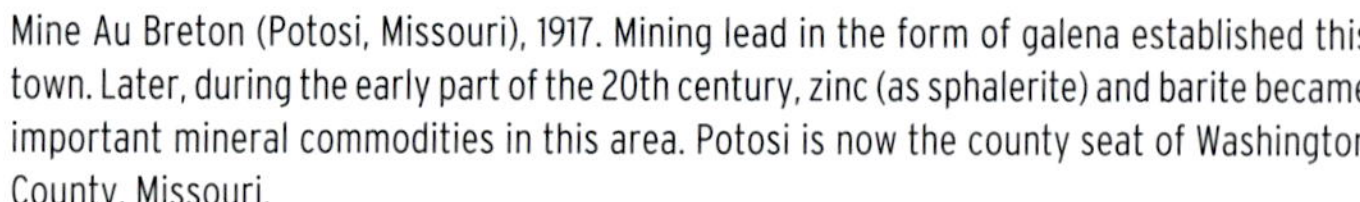

Mine Au Breton (Potosi, Missouri), 1917. Mining lead in the form of galena established this town. Later, during the early part of the 20th century, zinc (as sphalerite) and barite became important mineral commodities in this area. Potosi is now the county seat of Washington County, Missouri.

Use of a hand windlass in the mining of both galena and barite, Washington County, Missouri, lead district. This method of mining goes back into antiquity. The windlass brought clay, rock, and ore to the surface where it could be hand-separated, extracting the desired minerals from the lifted clay and rock. This method of mining continued in many parts of the world, including Washington County, until the 1950s. This painting and others showing barite mining was done by St. Louis (and Taos, New Mexico) artist O. E. Berninghaus, about 1915.

One of the first geologists to investigate MVT minerals from a modern geologic perspective was David Dale Owen (1805-1868). Owen started investigations in the 1830s under the United States General Land Office, where the mineral branch of that agency mapped geology in the U.S. Midwest. Its purpose was to geologically explore and determine within the Territories of Arkansas, Iowa, Wisconsin, and Minnesota those lands which should be reserved as mineral lands for mining and therefore not to be available for homesteading when these territories became states. Utilizing fleets of canoes launched at the mouths of small rivers, rivers which drain eastward into the Mississippi, Owen systematically investigated rocks exposed along these rivers, mapping geology and looking for evidence of mineralization in the rocks encountered. In his investigations, he identified a number of areas where lead and zinc mineralization occurred where these minerals were associated with calcite (spar), quartz crystals, and geodes. He considered all of these occurrences to be of igneous (hyperthermal or hydrothermal) origin and that they continued with depth and possibly became richer with greater depths. Lead and zinc minerals known and mined early in the 19th century came from the original "Tri-State-Region," the area where Illinois, Wisconsin, and Iowa came together (and now indicated by the place-name of Galena, Illinois). These minerals when later mined intensively in the late 19th century, were found to be mostly near surface deposits and appeared not to increase in richness with depth—or in other words, they did not appear to have a deep hydrothermal origin.

Hauling ore by team and wagon, Potosi, Washington County, Missouri, lead-barite district c. 1915.

Quartz vein characteristic of hydrothermal mineral deposits. Quartz like this was emplaced in the rock as a hot vapor, which might contain silver, gold, tungsten as well as other valuable metals. Quartz veins would be followed by prospectors over considerable distances searching for places where these valuable elements may have been emplaced. If such a place was found, they would expose the vein in what is known as a prospect.

Prospect, or small mine, developed along the previously shown quartz vein in an area where silver mineralization was discovered in the late 19th century. The site was excavated, but the amount of silver-bearing mineral found was not great enough to actually open a mine.

Porous chert characteristic of MVT deposits. Porous rock with lots of cavities like this can be the host rock with which MVT type mineralization is associated. This porous rock was once part of a fossil algae mass (stromatolite), a feature that is particularly conducive to MVT (and epithermal) mineralization. MVT mineralization normally does not continue with depth, unlike hydrothermal mineral deposits that usually do and are associated with quartz veins.

Silver mineralization from the above quartz vein and silver prospect. The silver is associated with galena, but it's a type of galena not found associated with MVT mineralization, which is very low in silver. The silver content of this galena can reach almost ten percent; such galena is known as argentiferous (silver bearing) galena and is always associated with hydrothermal deposits.

Native gold in quartz. Deep hydrothermal deposits can contain not only silver, but also gold, where this noble metal, like silver, is found associated with quartz veins.

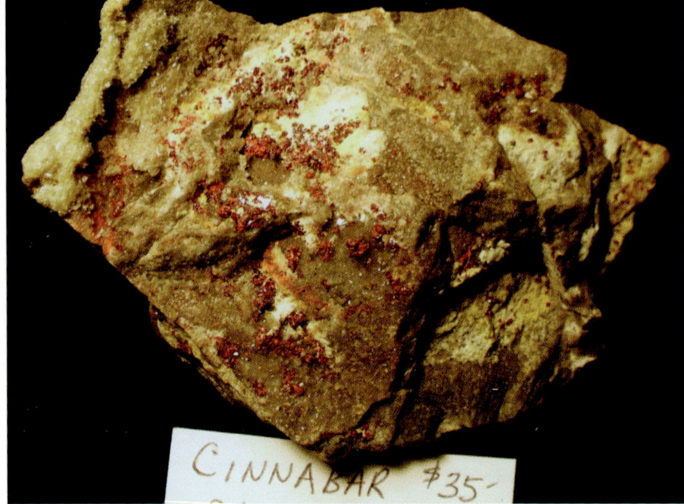

Cinnabar, a mercury mineral associated with shallow hydrothermal deposits. Silver, gold, and other precious metals generally are associated with what are known as deep-seated hydrothermal deposits, which formed fairly deep in the earth's crust and were exposed near the surface only by the removal of thick masses of overlying rock–overlying rock that might have had mineralization, like the cinnabar shown here, which came from southern Arkansas.

Tri-State Region Where Illinois, Wisconsin, and Iowa Converge

What is known as the northern Tri-State mining region was in the area of the junction of the states of Iowa, Wisconsin, and Illinois. This is in what is known as the driftless region, an area that was not covered by ice age (Pleistocene) glaciers and that also is a region of considerable relief. The ore found there consisted almost entirely of galena and sphalerite associated with calcite—limestone and dolostone of Middle Ordovician age being the host rock.

When extensive mining activity of MVT minerals began in the 20th century, it became even more apparent that deposits like those described by David Owen (Owen was most familiar with the Upper Mississippi northern area) did not extend with depth and in fact many of them were richest near the surface and seemed to peter out with depth. Also, more intense geologic exploration failed to show any connection with igneous activity, a prerequisite for hydrothermal mineralization and a phenomenon that appeared to be minimal in the region. There was also a richness in mineralization in these deposits where the associated rock was especially porous. It was in the 1920s, with increased emphasis on petroleum exploration when it was discovered that MVT mineral occurrences were similar in many ways to the geologic occurrence of petroleum. MVT mineral deposits appeared to mimic the geologic occurrence

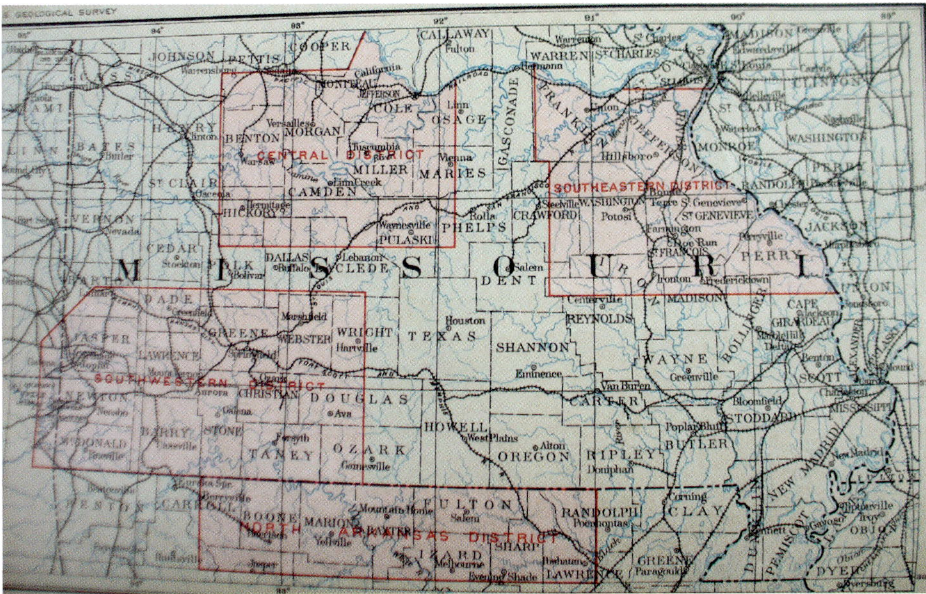

Generalized major mineral regions of the Ozarks (shown in pink) from Bain, Van Hise, and Adams, "Preliminary Report on the Lead and Zinc Deposits of the Ozark Region, U.S. Geological Survey, 1901." The Southeastern District includes the Washington County barite district, the Old lead belt of Flat River, and Bonneterre and the Mine La Motte-Fredericktown area. The Central District is primarily a barite region. The Southwestern District includes the Tri-State region, as well as lead and zinc occurrences to the east, like those near Springfield, Missouri (Pierson Creek). The North Arkansas District is primarily one of zinc minerals, which includes the mining area of Rush, Arkansas.

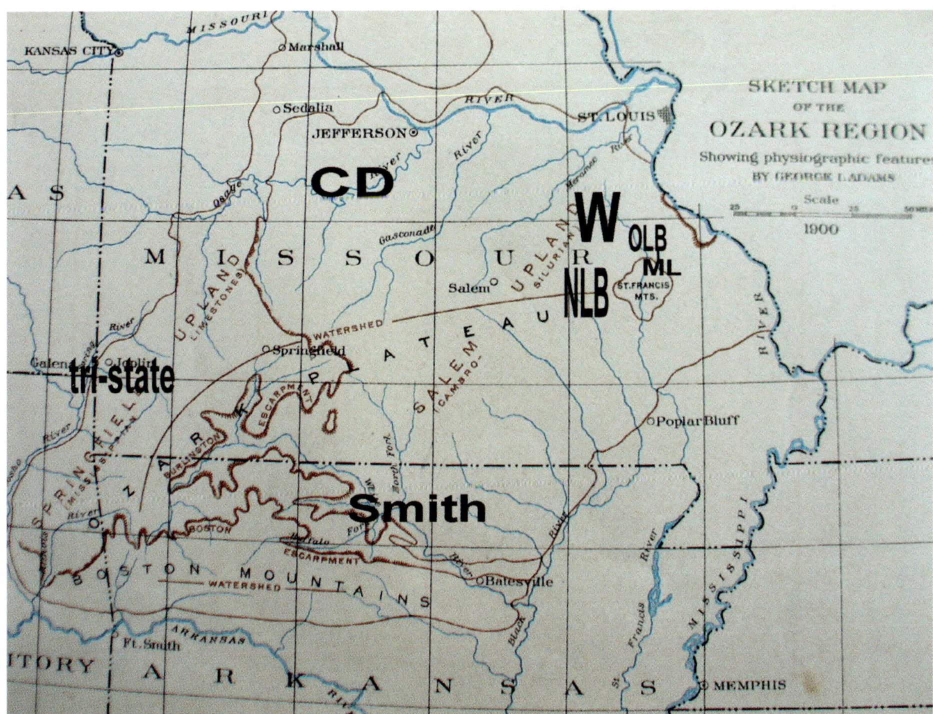

Sketch map of the Ozark Region. All in the Southeastern Region.
W is the Washington County barite area, OLB is the "old lead belt" of Flat River (Park Hills), ML is the Mine Lamotte and Fredericktown area, and NLB is the "New Lead Belt" or Viburnum Trend.
CD is the Central District.
Tri-State is the Joplin, Missouri; Picher, Oklahoma; and Galena, Kansas zinc area.
Smith is the zinc region, primarily smithsonite of northern Arkansas.

of petroleum then being found in the "oil fields" of the U.S. Midwest. This seemed to be reinforced also by the occurrence of asphalt being found at the time in the Tri-State mines, especially those mines that extended westward into what was more deeply buried strata under the states of Kansas and Oklahoma. There the mineral bearing strata tilt downward or dip westward because one is going off of the Ozark Uplift.

Midwest MVT Minerals

MVT mineral occurrences are associated with a limited number of mineral types, but minerals that do occur always include galena and sphalerite, with the lead-zinc occurrences found in Paleozoic carbonate rocks of the southern mid-continent being in fact the U.S. "type region" for MVT minerals. Galena associated with barite occurring in residual clays of the Ozark region—especially the Potosi-Washington County barite district of Missouri (Chapter 4)—is also considered a type of MVT deposit although mineralization is minimal in actual bedrock. Other barite occurrences similar to this, like the Cartersville Georgia barite district, can be considered as a logical extension of Missouri barite occurrences (they are both very similar). These types of occurrences, both of which are very close to the earth's surface and are associated with products of weathering, are known as extreme epithermal mineral deposits. The more deeply occurring ones found in bed rock, like those of the original Tri-State Illinois, Iowa, and Wisconsin mining region, are considered as epithermal mineral deposits of a different kind. Mineral deposits obviously associated with igneous activity and formed at great depth are known as hydrothermal—those intermediate between these two extremes are known as mesothermal mineral deposits. MVT-type mineral occurrences are usually considered as classic examples of epithermal deposits, but to some of them, like the Illinois fluorspar district or even parts of Missouri's Viburnum Trend, a mesothermal label might be applied.

PRELIMINARY REPORT ON THE LEAD AND ZINC DEPOSITS OF THE OZARK REGION.

INTRODUCTION.

By C. R. Van Hise.

The main purpose of this introductory chapter is to compare the succession of lithologic formations of the various lead and zinc districts of the Mississippi Valley, to summarize the salient points as to the occurrence of the ores common to all the districts, and, finally, to show the relation between the facts of the occurrence of the ores and the stratigraphy.

As I first applied the theory of two concentrations for the lead and zinc ores of the Mississippi Valley to the Wisconsin district, and as Mr. Bain does not treat this district at all in the present paper, I will first consider the Upper Mississippi Valley region.

ORES OF UPPER MISSISSIPPI VALLEY.

Following Chamberlin, I hold that the lead and zinc now composing the ores were widely dispersed through the Silurian limestones. I hold that the ores were first segregated in the limestones by an artesian circulation, confined above by the Cincinnati shale and below by the Trenton shales. Later the upper shale was removed by erosion, and from this time on the conditions were changed and the descending meteoric waters produced a second concentration. Thus were formed the ore deposits now mined. Professor Chamberlin does not recognize the first stage of concentration.

The following statements in reference to the Upper Mississippi Valley are largely compiled from my paper on Some Principles Controlling the Deposition of Ores:[1]

As a result of the most excellent work of the second Wisconsin survey under Chamberlin[2] and the later work of the Iowa survey,[3] we

[1] Some principles controlling the deposition of ores, by C. R. Van Hise: Trans. Am. Inst. Min. Eng., Vol. XXX, 1901, pp. 27-177.

[2] The ore deposits of southwestern Wisconsin, by T. C. Chamberlin: Geology of Wisconsin, Vol. IV, 1882, p. 412.

[3] Lead and zinc deposits of Iowa, by A. G. Leonard: Iowa Geol. Survey, Vol. IV, 1897, p. 23.

22 GEOL, PT 2—01——3 33

Cover page of "Preliminary Report on the Lead and Zinc Deposits of the Ozark Region" by Bain, Van Hise, and Adams, U.S. Geological Survey, 1901—one of the more extensive earlier works on MVT minerals and mineralization of the Ozark Region.

Minerals of the Southwestern district.

Name.	Composition.	Probable constitution.
Sphalerite (blende)	ZnS	ZnS.
Calamine	$H_2Zn_2SiO_5$	$2ZnOH_2OSiO_2$.
Smithsonite	$ZnCO_3$	$ZnCO_3$.
Hydrozincite		$ZnCO_32Zn(OH)_2$.
Goslarite	$H_{14}ZnSO_{11}$	$ZnSO_4+7H_2O$.
Galena	PbS	PbS.
Cerussite	$PbCO_3$	$PbCO_3$.
Anglesite	$PbSO_4$	$PbSO_4$.
Pyromorphite		$(PbCl)Pb_4Pb_3O_{12}$.
Leadhillite		$4PbOSO_32CO_2H_2O$.

[1] Kansas University Quarterly (A), IX, No. 2, pp. 161-165.

A selection of minerals of the southwestern Missouri (Tri-State) district. Some of these are also found in the other districts.

112 LEAD AND ZINC DEPOSITS OF OZARK REGION.

Minerals of the Southwestern district—Continued.

Name.	Composition.	Probable constitution.
Marcasite	FeS_2	FeS_2.
Pyrite	FeS_2	FeS_2.
Hematite	Fe_2O_3	Fe_2O_3.
Limonite	$H_6Fe_4O_9$	$2Fe_2O_3+3H_2O$.
Melanterite	$H_{14}Fe_2SO_{11}$	$Fe_2SO_4+7H_2O$.
Copiapite		$2Fe_2O_35SO_318H_2O$.
Chalcopyrite	$CuFeS_2$	$CuFeS_2$.
Malachite	$H_4Cu_2CO_5$	$CuCO_3Cu(OH)_2$.
Azurite	$H_4Cu_3CO_5$	$2CuCO_3Cu(OH)_2$.
Cuprite	Cu_2O	Cu_2O.
Covellite	CnS	CuS.
Chrysocolla		$CuSio_3+2H_2O$.
Aurichalcite		$2(ZnCu)CO_33(ZnCu)(OH)_2$.
Caledonite		$(PbCu)SO_4(PbCu)(OH)_2$.
Linarite		$(PbCu)SO_4(PbCu)(OH)_2$.
Greenockite	CdS	CdS.
Barite	$BaSO_4$	$BaSO_4$.
Gypsum	H_4CaSO_6	$CaSO_42H_2O$.
Calcite	$CaCO_3$	$CaCO_3$.
Dolomite	$CaMgC_2O_6$	$CaCO_3MgCO_3$.
Quartz	SiO_2	Completely crystalline silica.
Chert	SiO_2	Partly crystalline silica.
Pyrolusite	MnO_2	MnO_2.
Sulphur	S	S.
Bitumen		A group of organic compounds of considerable variety in composition.

Additional minerals of the southwestern district (From Bain, Van Hise, and Adams).

Sub-districts of Ozark MVT Mineral Deposits

Five sub-districts of MVT mineral deposits occur over the Ozark region. These are:

1. The Washington County, Missouri, galena and barite district, the earliest one to be mined and the southeastern Missouri lead belts consisting of the "Old Lead Belt" of St. Francois.
2. The Madison County and the Viburnum Trend (New Lead Belt) of Iron, Reynolds, and Shannon Counties.
3. The central Ozarks lead and barite district, occurring primarily just north of Lake of the Ozarks.
4. The southwestern Missouri Tri-State zinc and lead district, centered around Joplin, Missouri, Galena, Kansas, and Picher, Oklahoma.
5. The Arkansas zinc (smithsonite) district centered around Rush, Arkansas.

Headframes of zinc (sphalerite) and lead (Galena) mines in the southwestern Missouri, Tri-State region near Joplin, Missouri. These mines produced large amounts of lead and zinc, as well as smaller amounts of other elements from the late 19th century until about 1980. The mines were the source of many excellent mineral specimens that have been distributed worldwide among museums and collectors. MVT mineral deposits are well known for their large and often spectacular crystals.

Head frame of one of the mines in Missouri's Viburnum Trend, a large MVT mineralized area (new lead belt), which currently produces a major portion of the world's lead. The Viburnum Trend also produces beautiful mineral specimens representative of MVT mineralization.

Headframe of a small underground mine in the Ozarks, 1900. The next step above using a hand windless was to use horsepower to power a hoist. Numerous small mines like this dotted the Ozark landscape in the late 19th and early 20th centuries. Some of these produced unique, and sometimes beautiful, mineral specimens of which, unfortunately, few usually were saved. Early collectors, like the Boentes, were instrumental in collecting and preserving some of these.

Sphalerite in chert from the Joplin Tri-State district.

Typical mining town in its heyday: the early 1940s. Both the southwestern Missouri Tri-State mines and the mines of the old lead belt at Flat River were active producers at this time. As seen here, the town and mines were intimately associated, a characteristic of earlier mining districts usually not seen today.

Barite surface mine, Tiff, Missouri, 1956. Barite is a mineral that was extensively mined from residual material (red clay) in Washington County, Missouri. The barite deposits are different from other MVT mineral occurrences in that the barite is not found in bedrock; rather, it occurs in red clay above bedrock.

Headframe of one of the Tri-State (Missouri, Kansas, Oklahoma) mines and tailings piles in the 1940s. The Tri-State mining district was one of the largest MVT mining regions in the world. It produced from numerous mines vast amounts of lead (from galena) and zinc (from sphalerite) until the 1980s, when the last mine in Oklahoma closed.

MVT mineralization can be associated with fossil stromatolite reefs. Stromatolites are organic structures produced by the physiological activity of photosynthetic organisms, primarily blue-green algae or cyanobacteria. Many MVT mineral occurrences are associated with stromatolites. This is a diagram of a cross section through a group of stromatolites. This type of stromatolite group is known as digitate because it resembles fingers.

Slice through a portion of a digitate stromatolite "reef"from Magmont Mine in Missouri's Viburnum Trend. Notice the elongated "finger" at the left.

Mass of stromatolites preserved in chert (but now removed by solution). These masses of chert were recognized in the 1950s to have formed from what originally were stromatolite reefs. They frequently are seen associated with MVT mineralization. This mass (and many like it) occur in the previously shown barite surface mine.

Mass of bladed barite from the area where it was associated with the previously illustrated chert stromatolites. Large masses of barite like this are interesting–they make nice yard rocks but are a bit bulky for a collection.

Bladed barite, typical hand-sized specimens from the above illustrated locality.

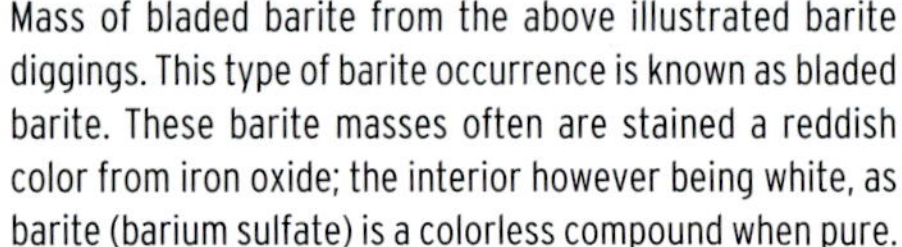

Mass of bladed barite from the above illustrated barite diggings. This type of barite occurrence is known as bladed barite. These barite masses often are stained a reddish color from iron oxide; the interior however being white, as barite (barium sulfate) is a colorless compound when pure.

Fast-flowing Eleven Point River.

The Eleven Point River, a spring-fed Ozark stream that flows through the Irish Wilderness. This area may harbor a large MVT mineral deposit at depth.

The Eleven Point River (Irish Wilderness) Region and MVT Mineralization

A controversial matter of mineral exploration developed in the 1980s regarding exploratory drilling in what is known as Missouri's Irish Wilderness, the least disturbed and wildest portion of the Missouri Ozarks. Through this region flows the Eleven Point River, a pristine and scenic river fed by numerous large springs. Geological data suggested that a sizeable ore body might underlie this region, an ore body, which besides the usual galena, might also be rich in copper, cobalt, and even nickel minerals—a reasonable scenario, as MVT mineralization generally becomes more complex toward the southern Ozarks. Eventually, the proposal to explore by drilling developed into an all-out "environmental battle" between mining and environmental interests. As the controversy ensued, environmentalists and aesthetic considerations ultimately won out, so it is still unclear whether a large mineralized zone might underlie the Irish Wilderness area, an area which, if large scale mining were developed, would to some degree be adversely affected by such activity.

Large springs like this one, known as Greer Spring, feed the Eleven Point River. They bubble up from water-filled conduits dissolved in the region's thick beds of dolostone.

MVT Minerals and Chemistry

There is usually a limited number of minerals found in MVT deposits. MVT deposits can also include minerals whose chemical and physical makeup differ from those of mineral occurrences formed at greater depths, such as those of hydrothermal origin. Regarding physical differences, MVT occurrences often include large, well-formed and excellent crystals—crystals formed in cavities, which are usually absent in the more deep-seated deposits associated with quartz veins. This is one of the reasons why MVT minerals are high on the list of geo-collectables. Chemically, the

Iron-oxide-stained, clear barite. This came from the Washington County barite district.

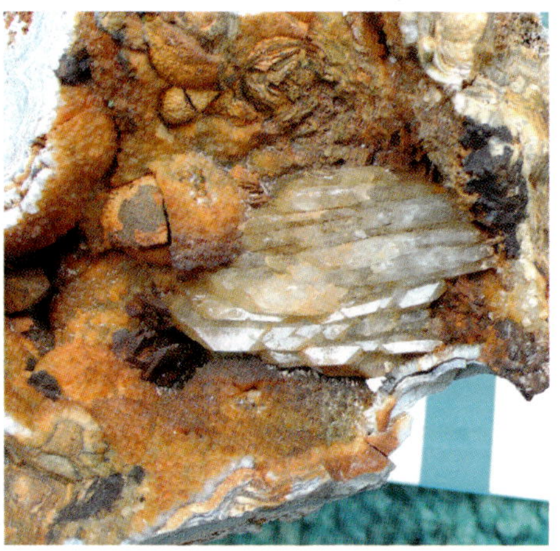

Clear barite associated with marcasite pseudomorphs. Washington County barite district. (Value range F.)

Fluorite. This attractive mineral is often associated with MVT deposits; however, it is also abundant in, and characteristic of, hydrothermal deposits.

Pyrite and calcite in mineralized vug, Arnold, Missouri. Vugs like this are fairly common in limestone and dolomite and can be lined with small crystals. They represent MVT mineralization on a small scale.

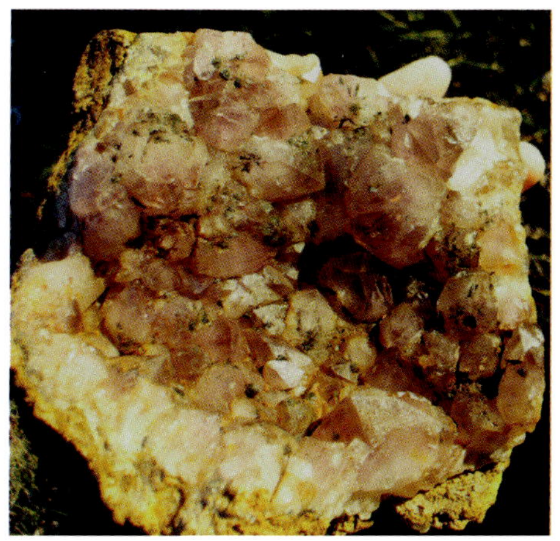

Amythest geode. These, at one time, were found in relative abundance associated with what are known as "filled sink" iron deposits of the Ozarks. They are associated with a peculiar type of mineral deposit somewhat unique to the Ozarks.

Geode. Geodes are commonly associated with limestone and dolostone of the U.S. Midwest. They represent a type of MVT mineral and crystal occurrence on a small scale.

most abundant metallic elements in MVT deposits are lead, zinc, iron, and copper. Minor metallic elements present can be cobalt, cadmium, strontium, barium, silver, and vanadium. Non-metallic elements that occur besides sulfur are fluorine and germanium. The abundance of lead, zinc, and to a lesser degree, copper is difficult to explain. Other metallic elements like tin, antimony, and molybdenum exist in stratiform mineral deposits in other parts of the globe, but these are either rare or absent in those of the western hemisphere. There appears to be a preference for specific elements, like lead and zinc in the northern hemisphere's stratiform deposits, while in other parts of the globe, as in parts of Asia, elements like tin, arsenic, and antimony take the place of lead and zinc in similar deposits.

Mineral Habit

Minerals are chemical compounds; as such they have very specific and distinct properties—properties characteristic of chemical compounds like color, luster, hardness, cleavage types, crystal structure, and shape. This also is the case with phenomena associated with minerals, such as electrical and magnetic properties. With minerals being products of nature, it might be expected that their properties would vary somewhat, but that variation is within a fixed set of parameters for any specific mineral. This variation, at least in part, often is a consequence of the presence of trace elements—trace elements being very small amounts of an element present in a mineral. Trace elements can influence a crystal's shape as well as other properties. An example of trace elements doing this is found with galena crystals, which contain only a small amount of silver (the silver atom can substitute for lead in galena's crystal lattice), a situation that can result in well-formed cubic crystals. The presence of slightly larger amounts of silver in galena's crystalline lattice will produce crystals of an octahedral shape and yet greater amounts will result in poorly defined crystals of any type. Galena crystals, as well as most other crystal types, usually will exhibit different and recognizable types of crystals, often ones that reflect their coming from a specific deposit. Such a variance in crystal form, a variance that can be dependent upon the locality where the crystal was formed, is referred to as crystal habit. Crystals of quartz, calcite, and other minerals sometimes can be identified not only as to the region from which they originated, but also (by a knowledgeable person using crystal habit) from what mine or even what part of a mine or outcropping the crystal originated. The phenomena of crystal-habit, as noted prior, is explained (at least in part) by the existence of trace elements in the crystal. Its widespread occurrence still is somewhat of a puzzle, however, as minerals, being chemical compounds, should always present the same crystal shape and crystal habit appears to contradict this rule in many cases.

Mining and Mineral Collecting

Mineral collecting is a time-honored activity that recognizes both the beauty and uniqueness of what sometimes are referred to as the "flowers" of the mineral kingdom.

Mineral specimens often are the only tangible things left from a former mine or mining region. Mineral collectors and collections preserve the history of a mining region through specimens as well as with related artifacts, like photos and even mining equipment. A mindset, unfortunately, sometimes exists in parts of the mineral industry, which almost seems to loathe geologic collecting. An example of such an attitude is found with mining companies that have implemented a policy of firing "on the spot" miners or other employees found collecting specimens, or who are caught taking specimens out in their lunch box. Advocates of such a policy claim it is necessary to prevent employees from collecting minerals while on company time. It should be pointed out that as ore, the value of minerals taken by miners is usually negligible and that such collecting usually takes place on lunch breaks. Such a hard-line policy effectively encourages a confrontational working environment between employees and employer. A similar attitude can exist toward the mineral collecting community in its refusal to even consider the collecting of mineral specimens and actively discouraging collecting as a company policy. Such a policy is not only a lost opportunity in public relations for the company itself, but again, it promotes a confrontational environment with a group of people who otherwise would be sympathetic toward and interested in the mining industry. The author was once told that "this is what public relations departments in large corporations are for" and a minority interested in minerals and mining history is considered nothing but an "inconsequential nuisance." Such attitudes have been prevalent at times with some U.S. mining companies; however, a more enlightened policy has more recently appeared, where programs of collecting choice mineral specimens through a concession and then selling them to collectors is implemented, thus preventing these jewels of the mineral kingdom from going into the rock crusher.

Pyrite in a calcite-lined vug, a common occurrence in Paleozoic limestone.

Mineral Collections as Historical Documents of Mining History and Mine Reclamation

A well-organized and curated mineral collection is more than a geologic and esthetic assemblage of specimens. Mining and mineral production can be either from surface excavations (as has often been the case with MVT occurrences) or underground mining. Of the latter, mining may go below the water table, requiring the use of pumps to keep water out of the working mine. When mining activity ceases, the mine may then fill with water. Surface mining, once mining activity has ceased, may require reclamation of the mined area to prevent surface water pollution and this obviously is an important consideration. More recently in the U.S., however, reclamation has often meant the removal of almost all surface indication of mining activity, with the land being required to be contoured as close as possible to its original configuration—often a very expensive proposition. Today, in many cases, minerals collected from a mine may be one of the few tangible documents of former mining activity. Sometimes, it seems that some members of the mining industry want to erase all evidence of past mining activity. Such a mindset usually comes from efforts at mine reclamation, which can be a good thing if not taken too far. Unique geologic phenomena can be exposed during mining and little consideration of its scientific value is sometimes given and a scientific treasure thus is lost. Mine reclamation is desirable; however, mining activity can also create many desirable features, such as depressions, which become bodies of water useful to wildlife, as well as interesting landscape changes and rock exposures that can have geologic interest and significance. Mineral and rock specimens may be all that remain of mining activities in these areas of geologic interest when reclamation has leveled everything, covers all rock exposures with soil, and regards everything associated with mining as "evil," or at least as politically incorrect. As a consequence of this, as a historical mining document, a mineral collection should have a high level of accurate information with its localities and other pertinent information accurately stated. Mine level information is desirable, but most important is the locality from whence a specimen came.

Older Collections and Mineral Specimens as Historical and Scientific Documents

Mineral occurrences usually are not obvious in nature with its natural outcrops along streams and other rock exposures. Mineral occurrences have to be discovered and the process of doing so is known as prospecting. In the main area of MVT minerals covered in this work, the Ozark Uplift of Missouri and Arkansas, such prospecting (which sometimes led to mining), took place throughout the 19th century. Unfortunately, minimal numbers of specimens at this time were collected and even fewer retained in collections. Mineral prospecting and mining continued into the 20th century with more specimens finding their way into collections, starting especially in the 1920s. Rockhound activity and geo-collecting really started in earnest after WWII, when the federal government gave monetary incentives to citizens to search for uranium. The author became introduced into the mineral and fossil scene in the 1950s where, as a boy, he met some of the persons who pioneered Ozark mineral collecting in the 1920s, a time when considerable amounts of prospecting for minerals were taking place. In this work, besides specimens from well-known localities like the Tri-State region of Missouri, will be seen mineral specimens that came from mineralized areas mined only on a small scale. Such occurrences are often of interest, as they exhibit habits different from those of the larger and better known mining regions.

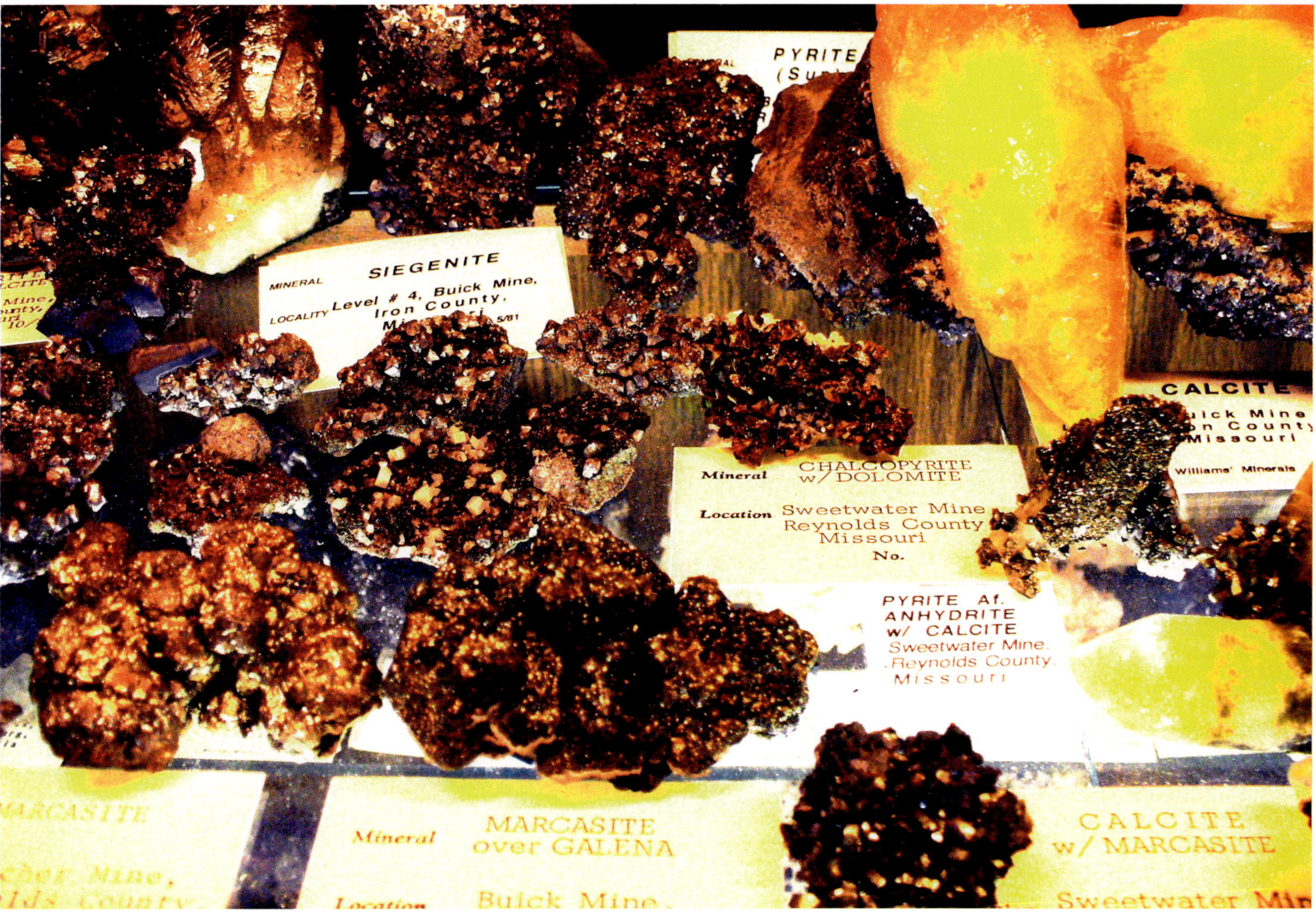

Display of MVT minerals from Missouri's Viburnum Trend. Courtesy of Glenn Williams.

One of the early collectors whose specimens grace this book are those of Ted and Elsie Boente, collectors who especially encouraged the author upon seeing their extensive and well-organized collection. They were also generous in sharing many field locations—locations that, at the time, were more accessible. The Boentes, in their basement museum in south St. Louis, had primarily Missouri and other Midwestern specimens, mostly self-collected. Later in life, Ted and Elsie worked with the St. Louis Museum of Science, where their collection of often rare and unusual Ozark minerals was given to the museum, which later became the St. Louis Science Center. Illustrated in this work are specimens from their collection, specimens from mineral localities that, otherwise, have largely been forgotten.

Value Range

MVT minerals are ones often seen at mineral, rock, and fossil shows. They are often collectable, as they can be attractive with their large and well-formed crystals. The following guide gives approximate values to specimens illustrated in this book.

Specimen Values

A	$1,000-$1,500
B	$500-$1,000
C	$250-$500
D	$100-$250
E	$50-$100
F	$25-$50
G	$10-$25
H	$1-$10

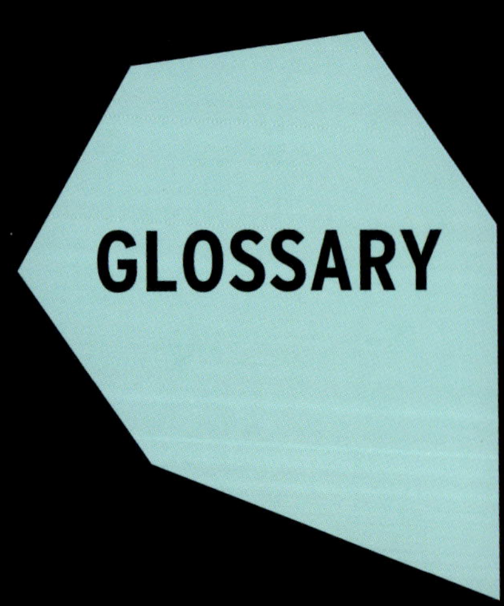

Carbonate rocks

Carbonate rock essentially consists of limestone and dolostone. Limestone is usually formed in shallow marine water from either chemical precipitated calcium carbonate or from the hard parts of marine invertebrate life like foraminifera, echinoderms, or mollusks.

Crystalline lattice

A three dimensional arrangement of atoms and molecules in a mineral or other solid which is responsible for the formation of crystals.

Dolostone

A chemically precipitated sedimentary rock composed of the mineral dolomite, which is calcium-magnesium carbonate. The term dolostone is used for the sedimentary rock composed of the mineral dolomite; it has physical properties similar to limestone.

Mississippi Valley Type (MVT) mineral deposits

Deposits of (often) heavy elements that occur primarily in limestone and dolostone of the southern mid-continental U.S. These deposits are mainly in the drainage basin of the Mississippi River, many being associated with the Ozark Uplift (Ozark Mountains) of southern Missouri and northern Arkansas.

Mineralization

The presence of (usually) economically valuable minerals in the earth's crust.

Periodic Table

A listing of the chemical elements according to their chemical properties and arranged in a manner showing repetitive chemical properties. This arrangement of elements also reflects the number of protons and neutrons (atomic number) in an atom's nucleus.

Pseudomorphs

"False forms," in reference to a mineral's crystalline structure. Pseudomorphs are minerals that have changed chemically for one reason or another, but still retain the crystal shape (or form) of the original mineral.

Stratiform Mineral Deposits

Mineral deposits found in sedimentary rocks where the mineralization is (more or less) parallel to the layers (strata) of these rocks. Stratiform mineral deposits are related to epithermal deposits—MVT deposits being both stratiform and epithermal.

Stromatolites

Organic or biogenic structures, which may be in the form of a series of domes or fingers (digitate stromatolites). Stromatolites usually are formed by the physiological activity of microscipic photosynthetic organisms, mainly blue-green algae or cyanobacteria growing in bodies of water.

Vug

A cavity or opening in rock. Vugs often are lined with crystals. Vugs in limestone and dolostone can contain, besides calcite and dolomite, crystals of MVT minerals. A vug can grade into a geode and commonly the two are often associated together.

Chapter One Resources

Bain, H. F., C. R. Van Hise, and G.I. Adams. 1901. *Preliminary Report on the Lead and Zinc Deposits of the Ozark Region.* U.S. Geological Survey, twenty second Annual Report. Washington, D.C.

Staebler, Gloria A. and Wendell E. Wilson, Eds., 2008. *American Mineral Treasures.* Lithographie, East Hampton, CT.

IT STARTED WITH LEAD

Galena and Other Lead Minerals

Economic interest in MVT minerals started with galena (a lead mineral) and to a lesser degree, sphalerite (a zinc mineral). Of the two, the lead minerals were of greatest interest, considering the wide variety of uses for this heavy, soft metal, especially in the production of armaments (bullets) and later in storage batteries. Zinc minerals were also important, zinc being used in the making of brass and other alloys. In the late 19th century, uses for zinc increased dramatically, with zinc plating (galvanizing) and electrical batteries becoming the foremost consumer of the element. In Europe, both galena and sphalerite were known to be associated with limestone as early as the middle ages—zinc minerals usually being associated and mined along with galena, but having little value until the late 18th century.

Galena, an Important Source of Lead

The mining of galena in the Ozarks started in the early 18th century (c. 1720) in the French Territory of Upper Louisiana. This was one of the earliest mining activities (by Europeans) in the New World. Note that the above states by Europeans, as Native Americans (Indians) had been mining flint, galena, mica, and

Galena (lead sulfide) crystals from the Tri-State (Missouri, Kansas, Oklahoma) mining district. Well-formed crystals like this are relatively rare. MVT mineral deposits produce some of the most spectacular mineral specimens in the world. (Value range D.)

Headframe of one of the smaller mines in Missouri's Viburnum Trend.

Disseminated galena. This is a typical sample of lead ore—fairly rich ore at that. Single crystals as shown in the previous photo are the exception rather than the rule with this mineral, as well as with most other minerals. Ore minerals usually consist of small crystals scattered through rock. Such is the case with MVT galena being disseminated through either limestone or dolostone.

Headframe of one of the larger mines in Missouri's Viburnum Trend, a north-south galena-rich mineralized region that occurs on the west side of a Precambrian massif known as the St. Francois Mountains.

Galena crystal with small octahedral galena crystals perched upon a large galena cube. Joplin, Tri-State Missouri area.

Galena crystal in barite. Much of the galena mined in the 18th and early 19th centuries was dug by hand and associated, as this is, with barite and red clay. Specimen from Tiff, Washington County, Missouri.

Mine Au Breton, late 1700s. The mining of lead was one of the earliest activities pursued by European settlers in the New World. By the late 18th century, the digging and smelting of galena became the basis for many small towns like this. Mine Au Breton later, in the early 19th century, was renamed Potosi, Missouri.

Typical small galena crystal—this one is from the central district near Lake of the Ozarks.

Close-up of galena crystal in barite from Washington County, Missouri.

obsidian for centuries prior to European settlement. French voyagers probably were informed by Native Americans as to how the galena occurred in a part of Upper Louisiana, an area that eventually would become Washington County in the state of Missouri, after the Louisiana Purchase in 1803. (It might be noted that the interior of North America in the late 1600s and 1700s was generally accessed from New Orleans). New Orleans was the capital of French North American territory—territory that extended northward onto the vast Canadian Shield, which, even at the time, was being explored by French trappers and other voyagers using the waterways, including the Mississippi River drainage basin.

Lead Mining in Louisiana Territory

The early method of lead mining consisted of digging into the soil until red clay (usually containing many rocks) was reached. Used in this method was a mechanism known as a windlass, an

Potosi, Missouri (formerly Mine Au Breton), 1917. The mining of lead and later barite were major economic activities in this area in the early part of the 20th century.

Rocker box. Galena (and later barite), after the clay on it had dried, was placed in a box on which were attached runners. The box was rocked back and forth to remove the red clay, which often stuck tenaciously to the desired minerals.

Large galena crystal with surface weathered and converted to anglesite. This was the lead ore sought in digging into red clay and bringing up clay, rock, and desired minerals, like galena (and later barite). Until the 20th century, only the (relatively uncommon) galena was a valuable commodity.

Weathered galena crystal converted to anglesite from the Potosi district. A considerable amount of anglesite has formed on the crystal showing cleavage directions developed in the underlying galena.

apparatus consisting of a log mounted on a log frame to which a crank was attached at one end with a rope and bucket at its center. Digging would continue into the red clay, with the excavated material being hoisted up by the rope and bucket. As this process deepened, barytes (barite, see Chapter Four) would appear along with rocks and red clay. As the digging went deeper into the red clay previously dug, material was placed around the circumference of the hole. Eventually, such digging might go to depths of over fifty feet into the red clay. As bedrock was approached, the amount of barytes present would increase and galena accompanying it would be found either as large anglesite-coated galena crystals or as shiny galena crystals embedded in masses of barytes. Material brought up was allowed to dry and a surface worker would put the clay-coated galena in a "rocker box," which, when rocked, would dislodge the clay from the galena. The surface worker would also remove galena from the barytes by cobbing and then placing it into a separate pile along with the anglesite-covered, weathered galena masses. This was the lead ore, which could now be smelted in wood-fired smelting operations.

Use of a windlass to bring up galena (and later barite) from below the surface. This method of mining persisted for over two centuries in various parts of the Ozarks. It was hard work and labor intensive, as the desired minerals were associated with large amounts of red clay and rock, which had to be separated from the desired minerals. Mechanized mining began in the 1920s and replaced labor-intensive hand-digging.

Weathered galena crystals from a mid-19th century lead mine. Galena, found associated with old lead workings, is usually covered with lead sulfate (the mineral anglesite) Unlike fresh galena, anglesite is ugly and, as such, is usually not a collectable mineral. Specimen at the bottom right has been cleaved to show shiny galena cleavage faces of a freshly broken crystal.

Galena in barite with associated limonite (originally marcasite) from the Washington County district.

Galena cube. An especially nice galena crystal from the Viburnum Trend, a site of the most recent discovery of large MVT galena occurrences, in the Ozarks of Missouri.

Lead Smelting in Upper Louisiana Territory

Smelting of galena was originally done with improvised wood-fired smelters in which much of the lead in the galena was lost and not recovered. These inefficient operations were carried out over a sizeable area of what is now Washington County, with areas around Mine Au Breton (Potosi, Missouri) being the earliest. Today, there are patches of ground that are contaminated with lead never recovered from these primitive smelting operations—a "toxic waste" legacy from the 1700s and early 1800s! As Washington County is still rural, gardens grown on this lead-laced soil may still result in lead-laced tomatoes or other produce—not too good! The presence of this lead has also inadvertently affected some of the groundwater. Parts of Washington County for this reason have recently (2011) become a focus for clean-up by the federal EPA (Environmental Protection Agency) superfund.

Mine LaMotte Lead Mines

What became a successful mining district (now near the town of Fredericktown, Missouri) was also discovered from surface galena occurrences in the mid-18th century. Originally mined as masses of galena occurring close to the surface in weathered dolomite, the Mine LaMotte deposits were found to extend to some depth as disseminated galena in massive dolomite of Cambrian age. In this district, associated with the galena later were found smaller amounts of the cobalt mineral siegenite, which occurs as small crystals associated with the galena.

Galena and siegenite in dolomite vugs. Associated with the Fredericktown and Mine LaMotte mineralization is the rare cobalt mineral siegenite. It is present as small crystals that can be seen in the middle of the photo.

Mine LaMotte galena. One of the characteristics of lead ore occurring in this area is the presence of numerous cavities (also known as vugs) in the ore bearing rock. These can be lined with dolomite crystals sometimes accompanied by galena cubes or octahedrons.

Small galena crystals perched upon dolomite. Few specimens of any size are seen today from the Mine LaMotte deposits that surround the Fredericktown, Missouri area. Most Mine LaMotte specimens occur as small crystals in vugs of crystalline dolostone (Bonneterre Formation of Cambrian age)

Transportation of Lead from the Upper Louisiana Mines

Lead (either as smelted lead or as galena) was the valuable commodity produced in Upper Louisiana. Outside of the Mine au Breton and Mine Lamotte areas, Upper Louisiana in the 1700s was a sparsely populated region. Lead being mined was transported to New Orleans and was then sent to Europe where it was used primarily in armaments. It's been said that much of the lead shot in hostilities between combatants in the 18th century (as between the French and the British) came from Missouri (or what would eventually become Missouri).

The Mississippi River was the major artery for transportation in the 18th century. Its tributary streams drained a large area that was unsettled (by Europeans) and was, to a major extent, under Native American control and occupation. Goods transported on these waterways included the lead mined in Upper Louisiana. Lead later would be transported eastward over land by wagon to the Mississippi, as was done in mid-19th century Missouri by the plank road, which became Missouri Highway 32. The method by which lead initially was transported was by the use of small boats on the waterways: waterways that flowed

The rough and rocky landscape of much of the Ozarks results in it being a region of clear, spring-fed streams. This is a typical view of one of the smaller streams. Such streams often served as "highways" in the 18th and early 19th centuries–today, they offer compelling and attractive recreation.

southward into the White River or into the Meramec River, which in turn empties into the Mississippi just south of what would become the city of St. Louis. The lead-producing region included areas west and north of Mine au Breton, areas where fast flowing, spring-fed streams occur. West of Mine au Breton, streams flow into the Courtois (pronounced *Court-a-way* in Missouri French), which in turn flows into the Huzzah and then empties into the Meramec. To the north was what would later be named the Mineral Fork River, which drains into what became known as Big River and in turn drains into the Meramec upstream from the Mississippi. Both streams would effectively carry wooden dug-out canoes loaded with either smelted lead or galena. This method of commodity transport continued after the 1803 Louisiana Purchase well into the 19th century, even after Missouri became a state.

Here an Ozark stream (the Little St. Francois River) has cut into the bottom-most beds of Paleozoic strata, the Lamotte Sandstone. Above this sandstone (now removed by erosion) was the Bonneterre Dolomite, the host rock for galena mineralization in both the old and new lead belts.

Placid waters of a previous river channel on the Meramec. The Meramec and its tributaries were used to transport minerals like galena and hematite of which the latter was mined on a fairly large scale in mid-19th century. This view is near Maramec Springs, an area of considerable iron mining in the 1850s.

High bluffs of Lower Ordovician and Cambrian age dolostone are found on the Meramec River, as well as on other Ozark Rivers.

Substantial bridges like this were built in the 1920s and '30s. These again gave a boost to mineral exploration and local mining, which reached its peak in the 1920s. This flurry of local mineral exploration saw numerous small mines developed, some of which produced mineral specimens of a habit different from those of the larger Ozark mining districts, large deposits like those of the old lead belt and the Tri-State regions.

Two modes of transportation! This bridge on the Meramec, just downstream from the previous photo, was constructed in the late 19th century. When roads were constructed in the Ozark region in the mid-19th century, rivers, like the Meramec, posed significant barriers to land transport. Ferries were utilized to cross the larger rivers until the late 19th century when bridges like this one, were built. The building of such infrastructure allowed greater mobility into what previously were very remote regions. This mobility, in turn, promoted more systematic and intense mineral exploration and extraction.

Modern Consequences of Dug-out-canoe and Other Small River Shipping

The author, an accomplished canoeist, has traveled a few thousand miles on Ozark waterways, which he considers to be some of the most attractive and delightful rivers on the planet. A combination of clear, spring-fed water combined with an abundance of rocky stream channels keeps Ozark waters pristine. This is accompanied by summer water temperatures, which can vary from a chilly 55 degrees to a pleasant (for swimming) 75-80 degrees F.—but always with beautiful water. As can often be the case with delightful things, a sour note came in the 1950s when a major landowner challenged a group of canoeists traversing the Meramec River with trespassing. Calling the local sheriff, the canoeing party was arrested and the trespassing charge stuck in the county court. Challenging this verdict, the canoeists managed to take the matter to the state supreme court, where this higher court ruled in favor of the canoeists. This ruling (Elder vs. Delcour, 1952) was based upon the fact that Missouri's waterways, during the early years of statehood, were widely used to transport by

Gravel bar camping. Recreation on Ozark streams has become a significant industry in the Ozarks of Missouri and Arkansas. In Missouri, the Elder vs. Delcour court decision, based upon the prior use of streams in commerce, gave the state's citizens the right to use the waterways as they had in the past. Gravel bar camping on a long Ozark float trip can be delightful.

Occasional small waterfalls like this can (somewhat) impede river transportation, including float trips. As most Ozark rivers have cut into relatively soft sedimentary rocks, such natural obstructions are relatively uncommon.

The geologic center of the Ozark Uplift is a "core" of very hard, Precambrian igneous rock. When a down-cutting stream encounters this hard rock, it produces what are known as shut ins. These can be challenging to traverse, even in a small boat. Rivers where a lot of this hard rock occurs usually were avoided in river transport unless the waters were high (and then were dangerous to both cargo and life). When low, they usually became too laborious to navigate.

Traversing these extremely rocky areas through Precambrian igneous rock requires both good canoeing skills, patience, and hard work.

Mist rises from the Current River from the large temperature difference between water (35 degrees F.) and the air (0 degree F.). Early river navigators were careful, skillful, and knew the river being traversed–modern ones should likewise.

Big River, a major tributary of the Meramec runs through the old lead belt. Today this river is quite attractive through this mining region although Missouri's older river guides state otherwise. This is because of both cleanup activities and natures propensity to heal herself with time. In this regard, however, the consumption of bottom-dwelling fish from the river is advised against, as the stream's bottom contains mine tailings containing small amounts of lead–lead which can be incorporated in small amounts in fish, especially bottom-dwelling ones. Mine tailings can compose part of the fine sediment found along the river, as on the sand bank to the left of the front canoe.

Miners with an ore truck in the old lead belt, Flat River Missouri, 1950. Most of the mines in the old lead belt were owned by the St. Joseph Lead Co. that used electrically powered transport in its mines; this smaller mine used a truck. Few mineral specimens of interest came from these mines for various reasons.

Ozark river in winter. Because they generally are fed by springs, Ozark rivers usually don't freeze during the winter. Early transportation by river of minerals and other commodities often was a year-round activity. Doing this, however, required good boating skills, as a spill in sometimes sub-freezing temperatures like this could be fatal.

small boat articles of commerce, which included galena. The public had a "right of easement" to use the waterways up to the normal high water mark, which included gravel and sandbars that occur frequently (Elder vs. Delcour, 1952). This right to use the waterways came from the transport of minerals by small boats, as well as from the transport of timber by floating it—not only on the Meramec, but on other Ozark streams as well. Streams, many with nearby lead deposits like the White, Osage, Gasconade, Current, and their tributaries—that is all streams which flow off of the Ozark dome—also served as "highways."

Missouri's Old Lead Belt

Centered around the towns of Bonneterre, Flat River, and other small towns (now mostly consolidated into the city of Park Hills) was once the largest lead mining region in the world. Originally discovered in the form of near-surface lead deposits in what were known as crevice deposits (which incidentally yielded a variety of secondary lead and copper minerals, see Chapter Ten), most of the lead mined was as disseminated galena in dolostone bedrock. As this Bonneterre Dolomite usually had few cavities and vugs, good mineral specimens from this region are few. Early mining in the richer portions of the deposits was also at a time when little interest was given to mineral collecting. The St. Joseph Lead Company, which controlled most of the mining in the 20th century, also had a strict policy against miners bringing out specimens from the mines.

Cave (and mine) along Big River in the old lead belt. Lead mineralization is often associated with red residual clay. This cave contains a lot of this and was once mined for galena. Some of the earliest mining activity along these rivers was for galena mineralization, a portion of which was associated with caves and other solution structures.

Another mine-cave entrance in the old lead belt region.

Early geologic map (from D. D. Owen, 1853) showing the Upper Mississippi (Tri-State area: Iowa, Wisconsin, Illinois). The river shown at the top of the map is the Wisconsin. A poor state of geologic knowledge of much of this area at the time is indicated by the bands of strata shown near the top. This does not occur in reality, as was shown by later geologic mapping. Note also the town of Galena at the northwestern corner of Illinois.

Galena Mined from the Tri-State Driftless Area Where Illinois, Wisconsin and Iowa Converge

This mining area was developed early—though not as early as that in southeastern Missouri—but became well known early in the 19th century. The northern Tri-State mining region is the area where the states (or territories) of Iowa, Wisconsin, and Illinois came together. It's an area that was not covered by ice age (Pleistocene) glaciers and, also, it's an area of considerable relief, tributary streams having cut into thick ore-bearing limestone beds as they flow to the Mississippi. The ore itself, consisting almost entirely of galena and sphalerite, is associated with clay and calcite. Mineralization is in Middle Ordovician limestone and dolostone, primarily the Galena and Platteville formations, carbonate rocks that locally can contain nice fossils and which crop out extensively in the upper Mississippi Tri-State area.

Winter ground transportation in the rugged area of the lead mines of southeastern Wisconsin, c. 1900. This so-called driftless area is one of considerable relief. Today, it is a popular ski resort region because of this and the fact that it lies due west of the population centers of Milwaukee, Wisconsin, and Chicago, Illinois.

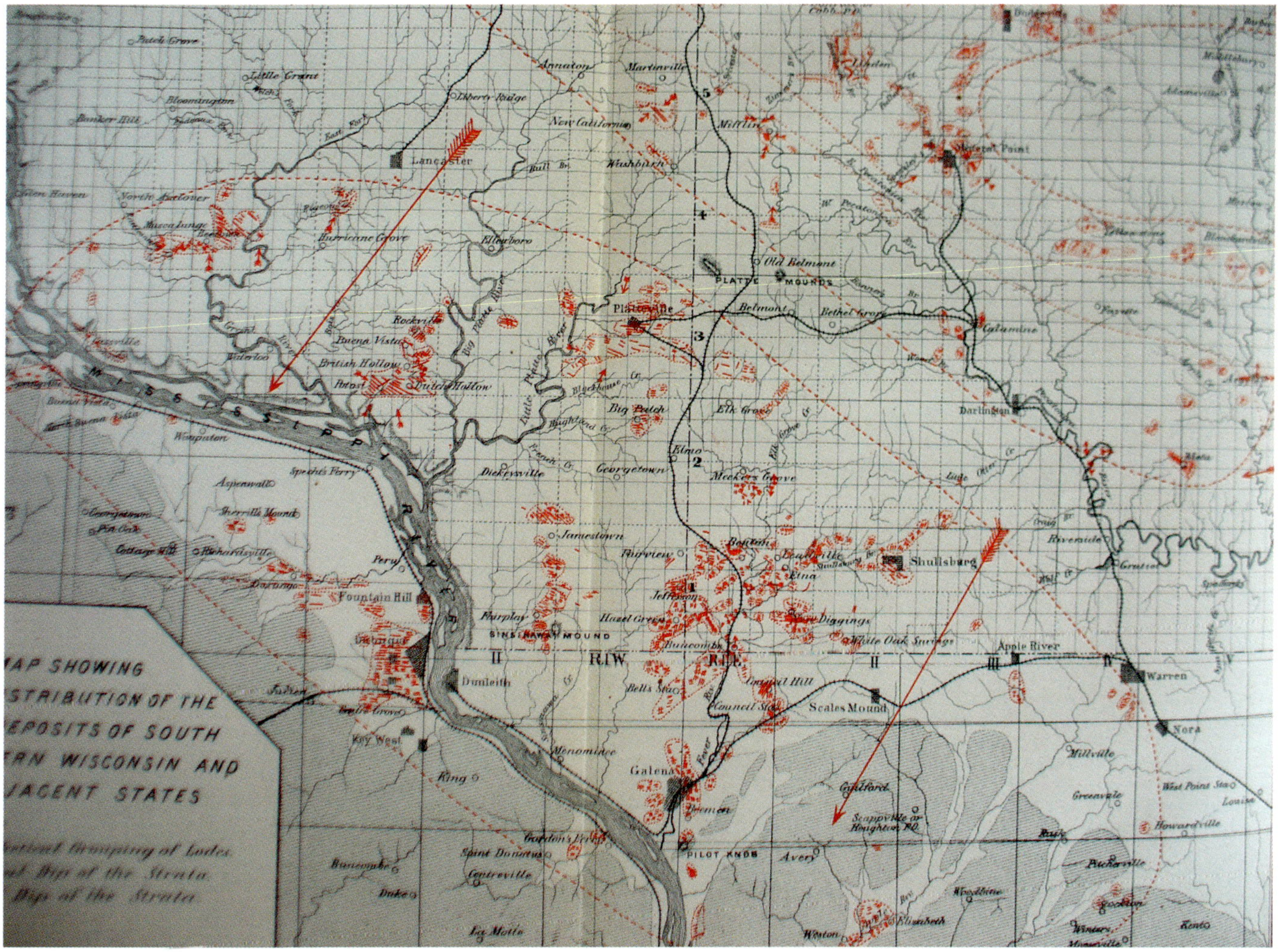

The upper Mississippi Tri-State mining region showing location of numerous mineralized areas in red. The Mississippi River is to the left, Illinois is at the bottom, and Iowa left of the Mississippi. This region produced large quantities of galena from about 1820 to 1960. Mineralization was associated with either the Galena Formation or the Platteville Limestone, both of Middle Ordovician age. From Bain, Van Hise, and Adams, 1901 (see Bibliography).

Calcite. As in most lead-zinc MVT deposits, calcite is common. Racine/Cuba City, Wisconsin. (Value range G.)

Cambrian strata (sandstone) exposed in high bluffs of the Wisconsin River, north of the mineralized area. These rocks underlie the galena-bearing Middle Ordovician rocks of the northern Tri-State mining region. This area, like the galena bearing regions to the south, is an area of high relief and was never covered by ice age glaciers; it is known as the driftless area of Wisconsin, N. Illinois, and eastern Iowa. From a 19th-century lithograph.

Galena from the northern Tri-State mining district. Galena was the predominant mineral mined in the district. Eagle Picher Mine, Schultzburg, Wisconsin. The Eagle Picher Mining Co. later became the dominant mining company in Missouri's Tri-State district. (Value range F.)

Sphalerite. Sphalerite was also produced in the Upper Mississippi Tri-State district. It, along with galena, were the two major commodity minerals mined from the district. Racine/Cuba City, Wisconsin.

Marcasite. Marcasite is a polymorph of pyrite; it is a frequent gangue mineral associated with lead-zinc primary mineralization. Racine/Cuba City, Wisconsin.

Stope working galena-rich ore. Dolostone of the Galena Formation is readily removed, especially where concentrations of galena and calcite were encountered. Note heavy shoring of the mine's ceiling; galena and other MVT mineralization is often associated with somewhat weathered and fractured rock, which may not form a strong and secure mine ceiling. Note the use of lanterns. The use of electric lights in mining began around 1905; mining in this region began in the early 19th century. This is an illustration from a late 19th century publication.

Mineralization in MVT deposits is often quite irregular. Here, a low-ceiling exploration tunnel connecting to galena bearing pockets is being traversed.

Tri-State-Mining Region (Missouri, Kansas, and Oklahoma)

Some of the most spectacular MVT minerals from the Ozark region come from what is known as the Tri-State district near Joplin, Missouri. Lead mining in southwestern Missouri began with the Pierson Creek mines near the James River, now near Springfield, Missouri, the large Tri-State deposits to the southwest being discovered later. Mining in the Tri-State district started in the late 19th century and continued until the 1980s, when the last mines closed. Since most of the mines were below the water table, most have now become water filled.

Open pit mine in the Tri-State area, Joplin, Missouri, 1900. Open pit mines such as this, or small underground workings, dotted the landscape in this mining region at the turn of the 19th century.

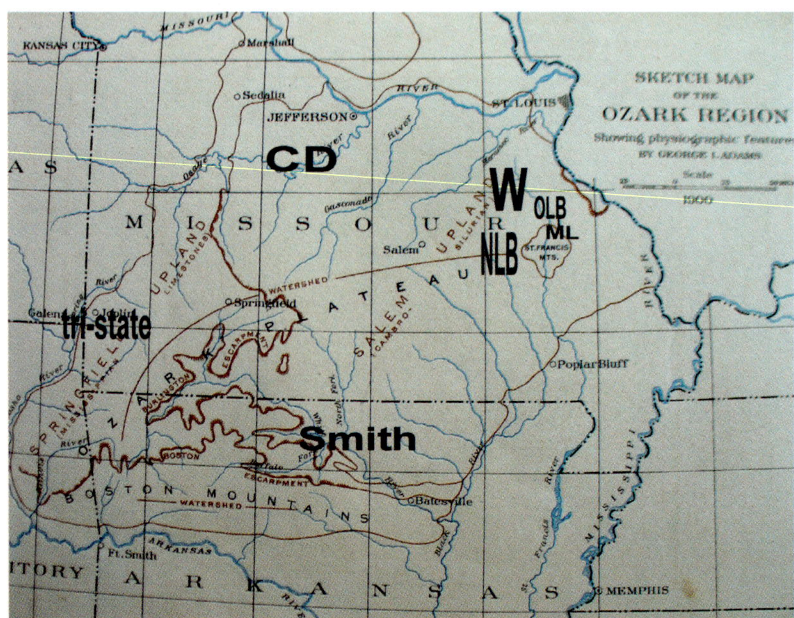

Major mineralized areas of the Ozarks.
The southwestern Missouri mining region is marked Tri-State.
CD is the central district of barite and galena.
Smith is the region of secondary zinc minerals (smithsonite) of northern Arkansas.
W is the Washington County barite and lead district of the northeastern Ozarks.
OLB is the "old lead belt" of Missouri, NLB is the "new lead belt."
ML is the Mine Lamotte mining region near Fredericktown, Missouri.

Ore stockpile and mill in the Joplin (Tri-State area) around 1900.

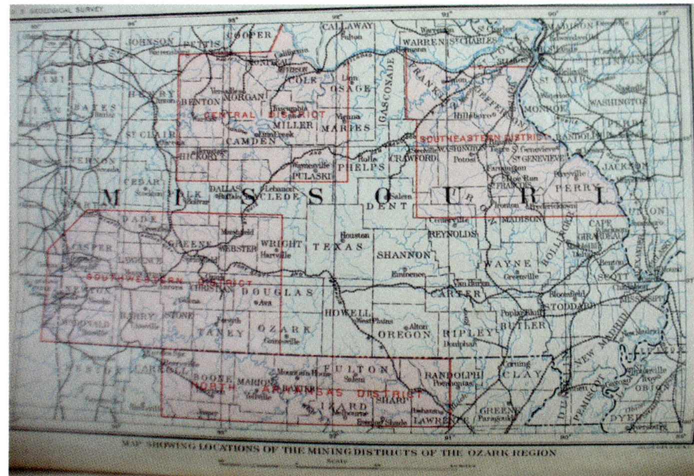

Location of mining districts of the Ozark Region (from Bain, Van, Hise, and Adams, 1901). The general extent of MVT mineralization was known by the end of the 19th century, but the western extent of the Joplin-Tri-State district into southeastern Kansas and what, at the time, was known as Indian Territory (now Oklahoma) was yet to be discovered. Also, the new lead belt (Viburnum Trend) would not be discovered until sixty years in the future. A large portion of the Viburnum Trend extends south of the red line into Reynolds County—a fact not known in 1901.

Headframe of small underground mine powered by a Missouri mule, 1900.

Cluster of galena crystals on sphalerite-covered chert. The presence of chert with galena and/or sphalerite is a good indicator that the specimen came from the Tri-State district near Joplin, Missouri.

The James River south of Springfield, Missouri. The Pierson Creek lead mines were located upstream from where this picture was taken. The James River runs into the White River (now Bull Shoals Lake), which flows into the Mississippi. These rivers were used to transport lead and lead ores, as was the Meramec and other Ozark rivers. Note the bridge, one built in the late 19th century and now replaced with a more sturdy one.

Galena and sphalerite, Pierson Creek Mines. The Pierson Creek mines were the first lead (and later zinc) mines in what would become southwestern Missouri. Some specimens from this old mining area can be quite attractive.

Reddish sphalerite crystals, known by the miners as "ruby jack," jack being a miner's term for the zinc mineral sphalerite. Ruby jack was often associated with the Pierson Creek mines, south of what is now Springfield, Missouri.

Galena and sphalerite on chert. Specimens like this came out of the Tri-State mines, where they were sold in rock shops along U.S. Highway 66, which passed through Joplin. (You go through St. Louie; Joplin, Missouri; Oklahoma City is mighty pretty–so the song says). When this song was popular in the late 1940s, many specimens were purveyed through local mineral entrepreneur Boodle Lang. This specimen, collected by the author, came by way of a later Tri-State mineral entrepreneur by the name of Chink Enders, who led somewhat illicit field trips where participants were taken to underground areas where they could extract specimens.

Weathered galena cube and crystal-filled vug from mine dumps in the Tri-State area. Until recently, many of the mine dumps in the Joplin area still produced such (somewhat) weathered specimens. Reclamation has more recently removed most of these spots.

Galena in chert, not on it. An unusual galena occurrence of lead and zinc, the honey yellow is from sphalerite. Note the impression in chert of a galena crystal at the right. Mindenmines, Lawrence County, southwestern Missouri.

Galena covered by cerrusite, Picher, Oklahoma. Cerrusite is a secondary alteration product of galena.

Galena octahedrons on a large galena cube.

Same specimen as above, different view. In the 1950s, specimens like this tempted tourists traveling along Highway 66 with its rock shops in the Tri-State area.

Viburnum Trend (Missouri's New Lead Belt)

The most recent MVT mining region in the Ozarks to be worked is known as the Viburnum Trend. The Viburnum Trend is a north-south mineralized region, which, in a way, is an extension of the Old lead belt where mineralization surrounds the St. Francois Mountains, an ancient area of exposed Precambrian (basement) rocks. This geologically ancient area of primarily igneous rock, and the topography carved from it, forms what is considered to be the geological center of the Ozarks.

Casteel Mine, 1960. Viburnum Trend, "new lead belt," Missouri. This relatively small mine in the Viburnum Trend has now been closed.

Ore processing mill near Boss, Missouri, in the Viburnum trend.

Working face with extensive galena mineralization (light-colored component) West Fork Mine, Viburnum Trend, 1980.

Aerial view of one of many mines in the Viburnum Trend surrounded by trees of Missouri's Mark Twain National Forest.

Magmont Mine, 1980.

Hand sample of high-grade lead ore, Magmont Mine.

Single well-formed galena cube. (Value range D.)

Calcite on galena cubes. A typical Viburnum trend specimen.

Galena cubes next to pyrite bars at middle top. These pyrite bars are one of the unique mineral occurrences of the Viburnum Trend; they came out of the Buick Mine in the 1970s. *Courtesy of Glenn Williams.*

More dodecahedrons from the West Fork Mine.

Dodecahedrons of galena. (Value range F.)

Display of Viburnum Trend minerals. Note the pyrite bars at the top. These are a mineral phenomena pretty well unique to the Viburnum Trend. *Courtesy of Glenn Williams.*

Galena and the Beginnings of Electronics

One of the novel uses for galena early in the last century was its use in what are known as crystal radio sets. In the late 19th century, with the development of wireless telegraphy, it was discovered that galena, a semi conductor, could act as a detector of radio waves. If a stiff steel wire connected with a radio circuit probed a galena crystal, and this radio circuit was connected to ear phones, Morse Code transmissions of wireless telegraphy could be heard. When it became possible to carry voice and music on radio waves in the early 1920s, the same arrangement of a steel wire (cat's whisker) probing a galena crystal enabled the public to effectively listen to the first radio broadcasts. This arrangement was simple, cheap, and effective—although not loud. Galena from Missouri usually was used in this arrangement and it became known as a "crystal radio" set because of the use of galena crystals. In the mid-1920s, audio amplification by use of the vacuum tube made the crystal radio obsolete.

A side issue relating to the above, which the author has heard about, is that in the 1930s through the 1950s, when lead miners in Flat River Missouri got small pieces of galena stuck in their teeth, they would sometimes complain of hearing voices and music, voices that were coming from nearby radio stations. Apparently, if close enough to a radio transmitter, a small galena fragment stuck in a tooth can act as a detector and produce a weak audio signal. I've also heard of dental fillings doing the same thing, so this probably is true. It might also be mentioned that up to the early 1950s, one of the Flat River stations broadcast a daily program in French—Missouri French. At this time, there were still some French-speaking persons living in the area west of Flat River, especially in the barite mining area of Washington County. This was a vestige of original French settlement—settlement which took place in this area well before the Louisiana Purchase.

Fossils Associated with Galena Deposits

The limestones and dolostones of the Upper Mississippi Tri-State area can contain relatively large and conspicuous fossils. The best known of these is *Receptaculites* and *Iscadites*, both coral-like fossils that are still paleontologically puzzling. Large cephalopods are also conspicuous and were sometimes associated with the galena mines.

The Viburnum Trend, as well as other lead mining areas in Missouri, produces its galena from geologically older carbonate rocks in which fossils are not so prevalent or as obvious as they are in the Upper Mississippi Tri-State area. Galena from southeastern Missouri comes primarily from the Bonneterre Formation, a thick dolostone sequence of lowermost Upper Cambrian age. The Bonneterre Formation in its easternmost outcrops locally can contain a rich assortment of trilobites—these are not, however, associated with the zones carrying galena. Molluscan fossils (planispiral gastropods

Crystal radio, 1923.

Marcasite and octahedral galena with calcite. (Value range F.)

Galena and pyrite. West Fork Mine. (Value range F.)

Galena and dolomite crystals. (Value range F.)

Cerussite. This secondary lead carbonate mineral was one of the ore minerals worked at the Granby mines. Few specimens were saved, however. Much more readily available today are specimens of this mineral from Morocco associated with galena, barite, and vandanite. (Chapter 10.) (Value range F.)

Cerussite crystal. This exceptionally large cerussite crystal is from MVT occurrences in Morocco. (Value range F.)

and monoplacophorans) do, however, occur where they are associated with stromatolite reefs and these fossils sometimes have been mineralized.

Secondary or Supergene Lead Minerals

Near-surface weathering and oxidization of sulfide minerals, like galena and sphalerite, can form what are referred to as secondary mineral deposits. These are deposits consisting of various sulfate and carbonate minerals (Chapter Five). In MVT deposits of the Ozarks, the most widely occurring secondary mineral is smithsonite—zinc carbonate, a mineral named after James Smithson of Smithsonian Institution fame. The best known MVT smithsonite occurrences are in northern Arkansas near the southern portion of the Ozark Uplift. Lead sulfate, the mineral anglesite, occurs also, but less commonly than does smithsonite and its occurrences usually are less attractive—sometimes being downright ugly, as is the case with that from Washington County. Other secondary MVT lead minerals are cerussite (a lead carbonate) and pyromorphite (a lead phosphate), both of which can be attractive minerals. The most extensive occurrences of these secondary lead and zinc MVT minerals were found during late 19th and early 20th century mining in the Tri-State district, especially in the vicinity of the town of Granby, a small town in southwestern Missouri. Here primary sulfide minerals weathered above the water table to form concentrations of these secondary minerals. Unfortunately, few specimens were saved from this district, but those that were can be quite attractive.

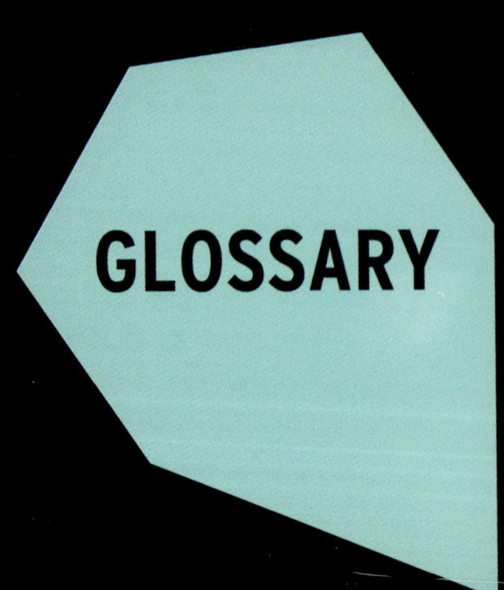

Cobbing

Breaking ore or other rock by hand to separate or remove various components. Galena was removed by cobbing from what, at the time, were worthless minerals collectively known as tiff and usually consisting of barite or calcite.

Superfund

A federal program through the EPA (Environmental Protection Agency) to clean up mining and industrial waste and pollution and funded by taxes on various potentially polluting industrial activities. Missouri's old lead belt has (and is) the location of a number of superfund cleanup activities.

Smelting

The process of converting a metallic ore into an industrially useful metal. Smelting is a chemical process involving heat, which produces molten metal. It is an ancient process known even in biblical times.

Louisiana Purchase

The transfer from France to the United States, in 1803, of a large portion of what would become the U.S. Midwest and west. After the purchase, various exploratory programs were instigated, the first being the Lewis and Clark expedition. Later, geological exploration, often on the areas rivers, was done to determine and map geology of the region for both general geological information, as well as determining the possible presence of ore bodies.

Chapter Two Resources

Buckley, Ernest R., 1909. *Geology of the Disseminated Lead Deposits of St. Francois and Washington Counties, Missouri*. Missouri Bureau of Geology and Mines, 2nd series, Vol. 9.

Currier, R., 1997. *Calcite and Marcasite Specimens from the Brushy Creek Mine*, Reynolds Co., Missouri. Rocks and Minerals, Vol. 72, pp 379-382.

Jones, Robert W., 1976. *The Viburnum Trend, Minerals from the New Lead Mines of Missouri*. Rock and Gem Magazine.

Keiser, H. D., 1930. *Mine La Motte; a historic lead property in southeast Missouri*. Engineering and Mining Journal, Vol. 130, No. 3, pp 110-114

Lasmanis, Raymond, 1989. *Galena from Mississippi-Valley-Type Deposits*. Rocks and Minerals, Vol. 72, pp 400-419

Lasmanis, Raymond, 1997. *Tri-State and Viburnum Trend Districts, an Overview*. Rocks and Minerals, pp 400-419.

McKnight, E. T. and R. P. Fisher, 1970. *Geology and ore deposits of the Picher Field—Oklahoma and Kansas*. U.S. Geological Survey Professional Paper, 588.

Medici, John C. and Jay Medici, 2008. Herkimer "Diamonds", Quartz from Upstate New York in American Mineral Teasures. Lithographie, LLC, East Hampton, Connecticut.

QUARTZ
Agate and Other Rockhound Delights

The most abundant mineral on the earth's surface is the oxide of silicon known as quartz! This hard mineral is well represented in MVT deposits, as might be expected, as well as in other stratiform deposits. Quartz, especially quartz of a colorful type, is also a favored material of rock hounds and lapidarists, and its wide variety of different types in Ozark occurrences has not gone unnoticed by them. The usual occurrence of quartz, which can be the host of MVT minerals, is a finely crystalline (cryptocrystalline) type in the form of chert and flint. Rounded surfaces covered by quartz crystals also occur, this being known as druzy (or drusy) quartz.

Quartz is usually not considered to be a MVT mineral; however, its frequent occurrence where it is intimately associated with classical MVT minerals, like sphalerite, galena, and barite, indicates that it was deposited along with these other minerals. Quartz associated with economic mineral deposits may differ little from that associated with residual quartz found on the surface. On the surface, such residual quartz, along with clay, represents a residue produced from weathering of limestone and dolomite of the same type as that occurring in the mines. Quartz crystals, therefore, found on the surface, like druzy quartz, is considered here as an MVT mineral, although one that may have no lead, zinc, copper, or cobalt mineralization associated with it.

Large quartz crystals, even in the lead and zinc mines where they might be expected to occur, are usually rare or absent. Rather, what is found are relatively small quartz crystals clustered on rounded surfaces and known as quartz druze (or druse).

Quartz druze is locally known in the Ozarks as mineral blossom, a quaint yet descriptive term. Some of these druzes can be thick and made up of multiple layers of different-colored quartz; thus, such material can be considered also as a type of agate. A coating of clear quartz druze is also associated with galena and marcasite in the Viburnum Trend, with some of these druzy coatings showing a directional trend, the quartz bearing fluids appearing to have been deposited from a specific direction.

Quartz is best known either as white quartz, which occurs in veins (quartz veins) like these, or as clear quartz crystals like those below from Arkansas.

Arkansas Quartz Crystals. These are probably the best-known examples of quartz crystals. They occur in the Ouachita Mountains of central Arkansas, which is a geologic province separate from the Ozarks—they also are not considered MVT minerals.

Druzy quartz with a stalactitic-like form of the botryoidal surface. Quartz druses are known for taking on many curious, fanciful, and peculiar shapes. (Value range F.)

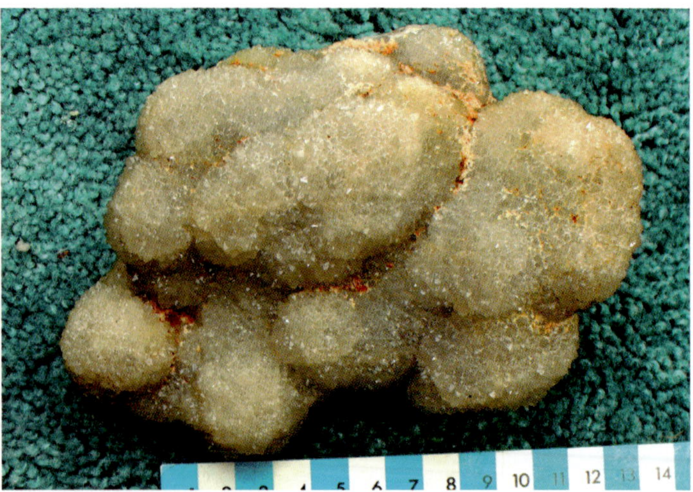

Druzy Quartz. Also known as mineral blossoms, these occur widely over the Ozarks, but are especially common in Washington County, as well as in other parts of the eastern Ozarks. Commonly, they exhibit what is known as a botryoidal surface—the type of surface seen here, as well as one seen on many other minerals. They usually are associated with red clay, particularly the residual red clay of the Cambrian Potosi Formation.

Clear quartz with pyrite, Casteel Mine, Viburnum Trend, Iron County, Missouri.

Drusy quartz associated with MVT lead mineralization. Druzy quartz, sometimes covered with sphalerite or galena crystals (or more likely covering chalcopyrite as seen at the left), can occur in the porous rock of Ozark lead and zinc mines. These specimens came from the Viburnum trend, new lead belt, Viburnum, Missouri. (Value range E.)

Druzy quartz coating chalcopyrite where the quartz-bearing solution originated from the left. Casteel Mine, Viburnum Trend. (Value range E.)

Druzy quartz coating chalcopyrite in a directional manner, a phenomenon typical of the Viburnum Trend. Casteel Mine, Viburnum Trend, Iron County, Missouri.

Back side of previously shown specimen.

Clear druzy quartz. The zinc mines of the Missouri Tri-State mining district (Missouri, Kansas, and Oklahoma) have produced clear druzy quartz specimens like this as do the mines of the Viburnum Trend. (This specimen came from an old collection, it needs to be cleaned.) (Value range F.)

Drusy quartz masses like this can be found scattered over the Ozarks. When it occurs in surface boulders, it often has a yellowish cast as seen here. Druzy quartz from mines, by contrast, usually is clear or even bluish in color. Surface specimen, Iron County, Missouri. (Value range G.)

A group of druzy quartz crystals from surface exposures, Washington County, Missouri barite district. (Value range F for group.)

Another quartz specimen from the Joplin, Missouri, Tri-State district. (Value range F.)

Clear quartz from the western portion of the Tri-State region (Oklahoma and Kansas) near Picher, Oklahoma. Minerals from the western portion of the Tri-State district often are more "gemmy" in appearance than are those from Missouri. (Value range F.)

Druzy quartz found on the inside of some surface rocks can be clear and lack the yellow iron staining. This clear quartz came from vugs in chert boulders of the Eminence Formation, Iron County, Missouri. (Value range G.)

Pink and orange druzy quartz. Ozark druzy quartz can come in a variety of colors. These small masses are pink and orange; others that occur over the Ozarks can be green, grey, black, and bluish.

Clear quartz covering sphalerite, western portion of Tri-State district, Picher, Oklahoma. (Value range F.)

Agates

Agates cut from quartz druze are referred to by rock hounds as Washington County agate, where it is especially common. The gem industry often despairingly calls such rock hound produced stones and jewelry "backyard jewelry" and some haughty persons regard it with contempt. But some of these agate cabochons not only represent very skilled work, but the quartz or agate itself is attractive and often uniquely positioned to make an attractive setting. The actuality of the matter is that many of the stones and jewelry made from these forms of quartz can have more character than do many of the precious stones, but to like such "jewels" is subjective—the attraction of a particular gemstone being a personal preference.

Three cabachons made from Washington County agate. The red band can be seen on the "cab" at the right. (Value range F for group.)

Associated with some Washington County agate is the pronounced red band seen in the middle of the picture. This red band comes from the mineral hematite and may represent a pulse of iron oxide mobilized and introduced by a large earthquake that occurred when the druze was forming. The time of formation of quartz druze is unknown; however, extensive seismic activity is evident over a large part of the Ozarks, particularly the eastern Ozarks. Such an earthquake may have originated millions of years ago along one of these faults or from sometime during the geologic past on the major seismic zone in southeastern Missouri where the historic New Madrid earthquake occurred in late 1811 and 1812.

Washington County agate. Druzy quartz can be associated with finely layered or banded quartz, which produces a characteristic form of agate often found in Washington County, Missouri.

The red band (see previous image) can better be seen here. These cabochons are made from typical Washington County agate. (Value range F.)

Union Road Agate and Geodes

Known as Union Road agates by rock hounds, these are agate concretions or nodules that occur in red residual clay (a weathering product formed from the solution of limestone). They are found (sometimes) in the same geological environment associated with MVT type minerals, where they occur without the galena and sphalerite so commonly associated with MVT type deposits.

Union Road Agate. Construction of I-55 south of St. Louis uncovered a zone of geode-like concretions with agate interiors. These entered the rockhound world and have become desirable geo-collectables. They probably can count as MVT minerals in a broad way, associated as they are with red clay weathered from limestone and, at times, also associated with barite.

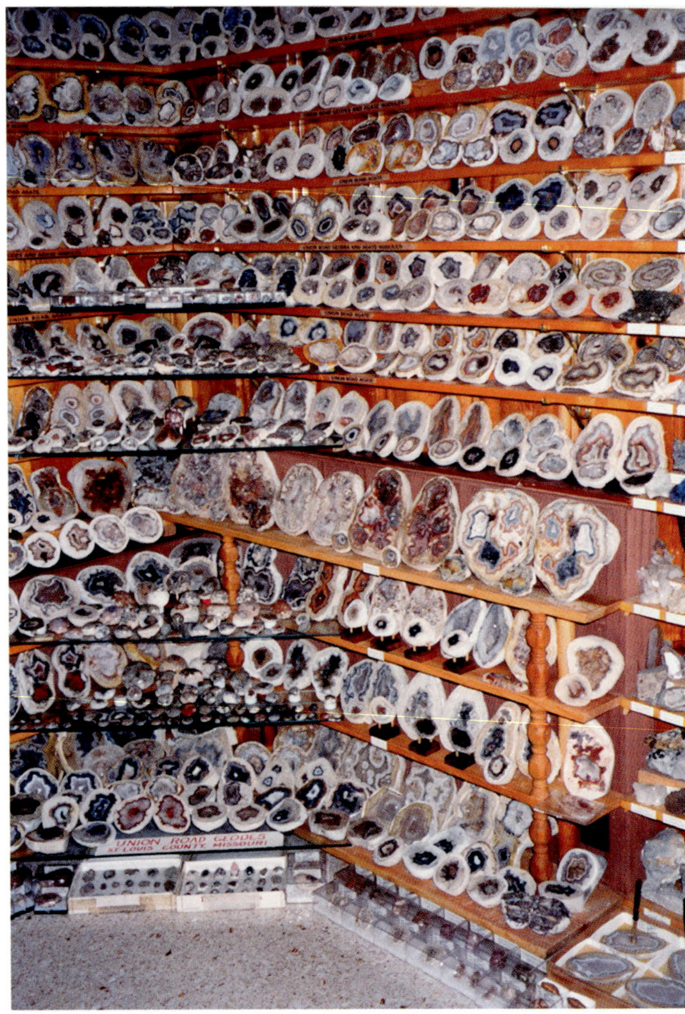

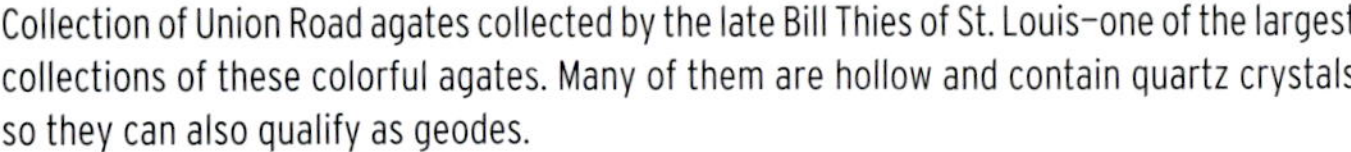

Collection of Union Road agates collected by the late Bill Thies of St. Louis—one of the largest collections of these colorful agates. Many of them are hollow and contain quartz crystals so they can also qualify as geodes.

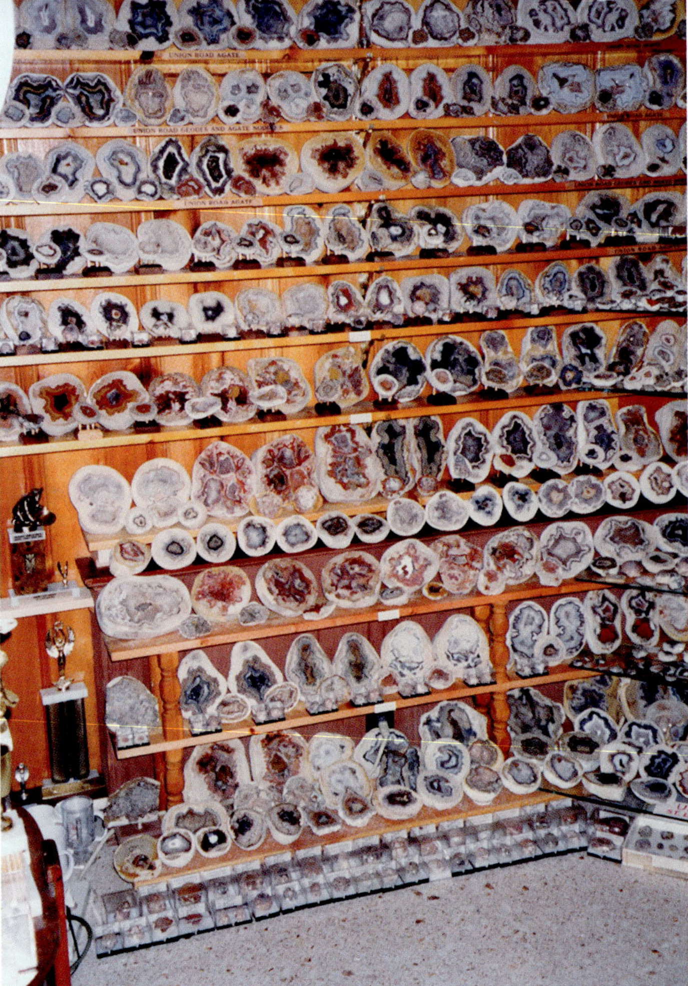

Another portion of the collection of Union Road agates in the Bill Thies collection.

Mozarkite and Jacks Jasper

Mozarkite (nothing to do with the composer) is a contraction of Missouri and Ozarks. This quartz lapidary material is also the Missouri state rock. It's found scattered over the Ozarks of Missouri and Arkansas in localized areas. Mozarkite is a type of flint; that is, it's a cryptocrystalline form of quartz containing a pink iron oxide pigment. Missouri was one of the first states to establish a state rock and mineral and mozarkite was chosen, as it is especially characteristic of the Missouri Ozarks. Mozarkite is attractive and it's a form of flint or chert—chert being a rock that is especially ubiquitous in the Ozark region of both Missouri

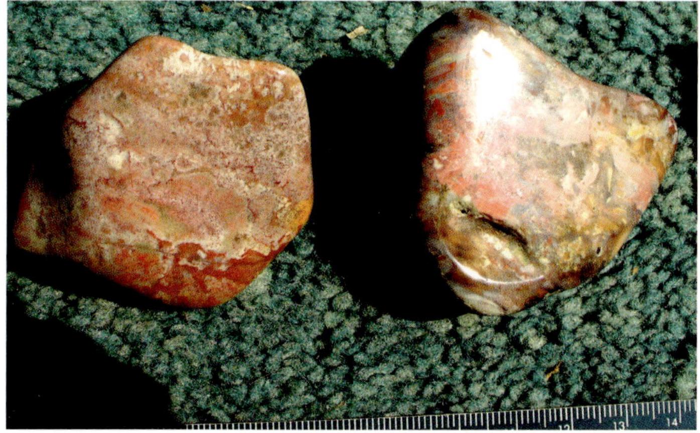

Polished pebbles of mozarkite, Lincoln, Missouri. Much of the mozarkite seen in rock shops has come from near Lincoln, Missouri, west of Lake of the Ozarks. (Value range G for single specimen.)

Mozarkite cabochon. Mozarkite is flint (a hard and glassy form of chert) containing a pink, ferric oxide pigment. This specimen contains small spheres known as oolites. Mozarkite is the Missouri state rock! (Value range G.)

Group of polished mozarkite from near Lincoln, Missouri.

and Arkansas. Missouri (and the Ozarks) might even be called the "chert (and flint) capital of the world," as this rock is especially abundant there. The name mozarkite is a rockhound term; it's a contraction of the words Missouri (*Mo*) and Ozarks with the mineralogical ending *ite*. Colored forms of chert, which occur over the Ozarks, were noted and its distinctive pink and purplish forms were given this name by rockhounds in the 1950s. As a teen, when mozarkite was designated the Missouri state rock, the author recalls some geologists being upset by this designation, as mozarkite was not to be found in any of the "official" literature of geology—the actual naming of mozarkite by the Missouri legislature as the state rock gave many of them heartburn.

Another quartz-rockhound related stone has to do with St. Louis's McDonnell Aircraft Rockhound club. McDonnell Aircraft (now part of Boeing Aircraft Co.), during the 1950s through the 1970s was one of the larger employers in the St. Louis area, their production of military aircraft at the time being supported by the "cold war." A lot of talented people took advantage of the company's generous support of company vocational organizations of which the Mac Rockhounds played an active part. Many skilled persons, often machinists who made aircraft parts—persons good with both their hands and minds—were attracted to the Mac Rockhound lapidary interests, wanting to work with something other than metal and quartz as a lapidary material really "filled the bill." Adventure associated with the discovery of quartz minerals in the Ozarks south of St. Louis was also appealing, so that a variety of forms of Ozark-related quartz surfaced and came into play as lapidary material. Besides mozarkite, there was Washington County agate, Jack's jasper and Union Road agate, to name a few.

Jacks Jasper. An iron oxide-rich chert breccia found in southeastern Missouri.

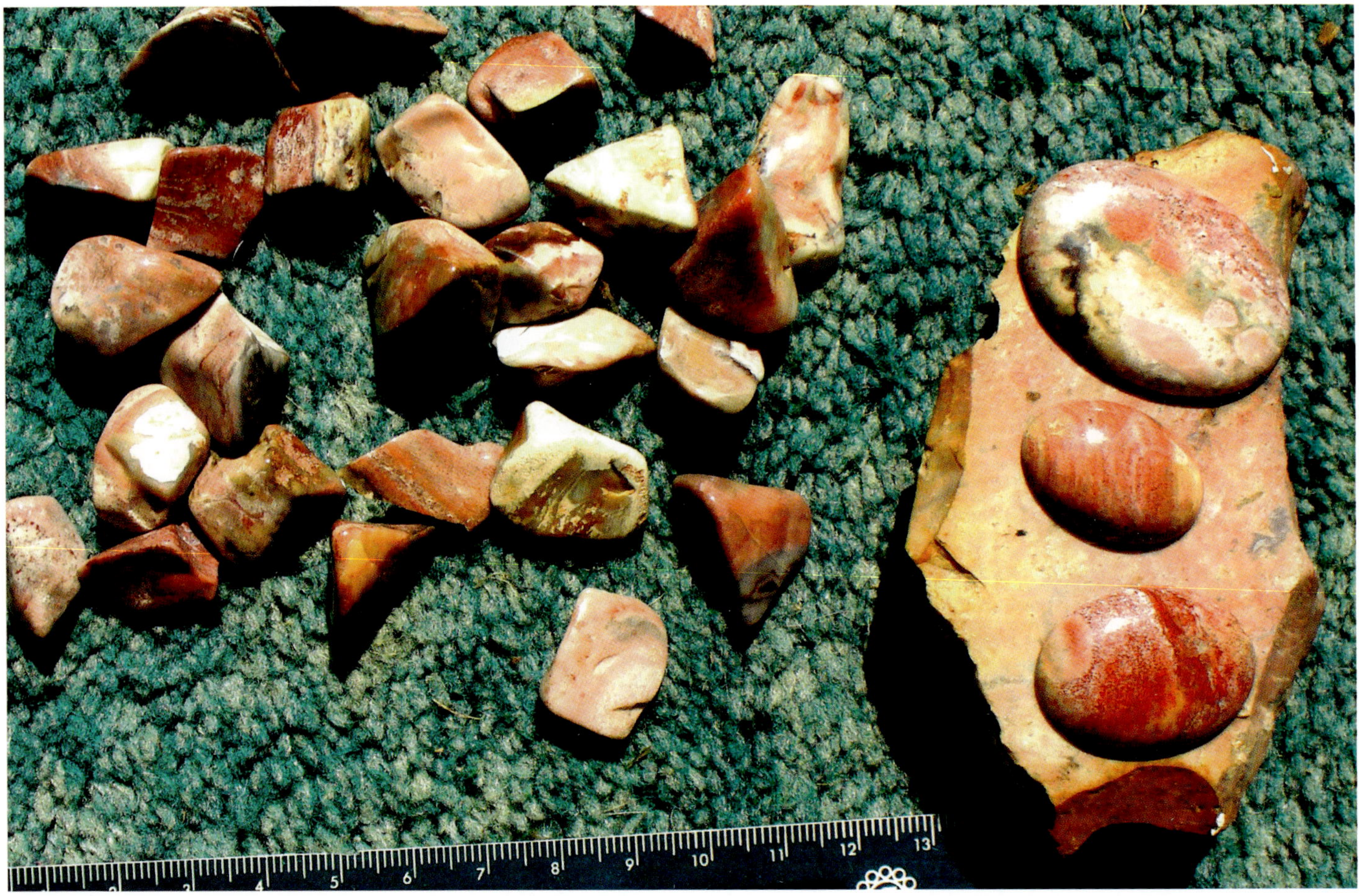

Polished pebbles of mozarkite (left) and cabochons (right) on a chunk of mozarkite.

Geodes

Geodes are peculiar, roundish quartz-rich rocks usually found in limestone or dolostone—they being especially characteristic of limestones of Mississippian age. Geodes occur widely in the Mississippi Valley region and, in a broad sense, can be considered as part of the MVT mineral scenario.

Geode, northeastern Missouri. A bewildering variety of quartz occurs in the interior of geodes–peculiar rounded rocks that are often hollow and contain crystals. Although not usually considered (in a strict sense) as an MVT mineral occurrence, geodes, in a way, are. They occur in the same limestone that produces MVT minerals and sometimes even contain sphalerite and galena. They also are often associated with limestone of Mississippian age and are found in the Tri-State area, sometimes associated with its mineralization.

Amethyst geode. Although farther from MVT mineralization than other geodes, amethyst-filled geodes were found in some quantity in the Ozarks where they were associated with what are known as filled-sink iron deposits. These peculiar deposits are found in the central and northern Ozarks; the origin of them has not been satisfactorily explained or well understood. Amethyst-filled geodes associated with these deposits are now quite rare and are little known to modern collectors, most of whom are familiar with amethyst geodes extracted from basalts in Brazil. This specimen came from the collection of Ted and Elsie Boente, mineral collectors who specialized in Missouri minerals and who also started collecting in the early 1920s. This was a time when it was easy to gain access to mineral (and fossil localities); however, only a few persons (like the Boentes) appreciated such things and understood their value at the time. *Courtesy of St. Louis Science Center.*

Amethystine geode, Rueppell Mine, Stanton, Missouri. (Value range D.)

Reddish amethystine crystals from Cherry Valley Iron Mine near Steelville, Missouri. (Value range F.)

Small amethystine quartz specimen, Cherry Valley Iron Mine.

Filled Sink Iron Deposits and Other Geologic Wonders and Curiosities

Arguably these are not really MVT deposits—they often are placed under their own category of filled sinks or mineralized sink holes but, as is the case with geodes, they can be shoe-horned under the MVT mineral category. Solution of carbonate rocks results in the forming of sizeable depressions—depressions that can then be filled with different types of materials. In ways not totally understood, some of these ancient sinkholes were filled with iron minerals, predominantly iron sulfide or iron oxide or both. Later, the sulfides became oxidized to hematite either entirely or partially and, if large enough, could form a mineable deposit. Iron minerals deposited in these "filled

Small amethyst geode from Moselle Mine No. 10, south of Rolla, Missouri.

Moselle Mine No. 10, a filled sink iron deposit, typical and somewhat unique to the Ozarks.

sinks" were extensively worked for both their iron and sulfur in the early part of the 20th century. Notable in these deposits were amethyst geodes, some of which were fairly large, attractive, and collectable—all quite different from the amethyst usually seen today, which come from cavities in basalt mined on a large scale in both Brazil and Uruguay.

Quartz replaced fossil corals from Florida. The interior of the coral has become a geode, which exhibits a botryoidal surface. These recently have become quite widespread on the geo-collectable market. Similar, but rare, botryoidal interiors of Paleozoic corals (Lithostronella) occur in the Mississippi Valley area. (Value range F for group.)

Snail-like fossil *Scaevogyra* in quartz druze. Fossils of rare and unusual mollusks are found associated with the peculiar druzy quartz or mineral blossom of the Ozarks. Tiff, Washington County, Missouri.

Amethyst druze, Magaliesberg Mountains, near Pretoria, South Africa. This quartz has appeared in quantity at the Tucson show; it's similar to druzy quartz occurrences in the Ozarks, except that Ozark druzy quartz rarely is amethystine. (Value range F.)

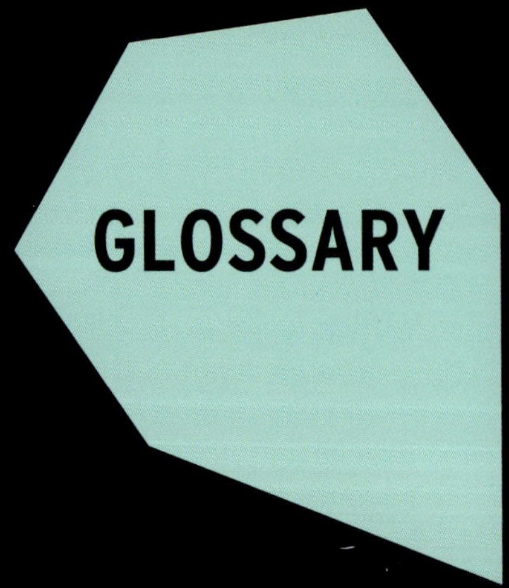

GLOSSARY

Commodity

In economics, something that is of value, as it serves society in significant ways and its value is widely recognized. Economic minerals like galena, hematite, and other minerals that are a source of industrial elements are a commodity. Quartz and rock associated with mineable minerals are not.

Density

A physical property having to do with the weight or mass of a mineral or other solid. Technically, density is weight per unit volume. High density minerals are noticeably heavy, like galena.

Gangue mineral

Minerals or material associated with a valuable mineral, but which in itself has no value as a commodity. Quartz is usually a gangue mineral.

Pegmatite

An occurrence of large crystals of rock forming minerals that are found in pockets or other limited masses and intimately associated with intrusive rocks like granite.

Residual (red) clay

Related to quartz druze and the Washington County barite district, this fine grained clay is left behind as a product or residue of chemical weathering and/or solution of rock. Sometimes this red clay is known as

Silicon

Silicon is element number 14 on the periodic table. It is in the same group as carbon being able to gain or loose electrons with equal ease; although what it usually does is to share electrons with another element. The oxide of silicon is quartz. Silicon is an example of a semi-conductor in electronics. It's a key player in the electronic "wizardry" of today where it reigns supreme at California's Silicon Valley.

Sulfophile element

An element that is attracted to sulfur and thus often occurs as a sulfide mineral. Lead, zinc, silver, and copper are some notable sulfophile elements—their occurrence often being as a sulfide where they are chemically combined with sulfur. Silicon is not a sulfophile element.

Vugs

Cavities or large holes in rock (often limestone or dolostone) Vugs can contain crystals including those of barite and quartz—sometimes crystals found in vugs can be especially nice and therefore collectable.

Chapter Three Resources

Medici, John C. and Jay Medici, 2008. Herkimer "Diamonds," *Quartz from Upstate New York in American Mineral Teasures*. Lithographie, LLC, East Hampton, Connecticut.

BARITE AND RELATED MINERALS

Barite or barytes, an archaic name for this heavy mineral, is a commonly occurring gangue mineral found associated with many mineral deposits. Barite has been known since ancient times, as it often was associated with minerals of value being mined at the time. Until the late 19th century, barite had little value as a commodity mineral. The old name for barite, barytes refers to its high density; it's a heavy mineral, especially for a light-colored, non-metallic one. The word barytes comes from the Greek word "bary," meaning "heavy or dense." One finds this root used today in particle physics, where heavy nuclear particles, like protons and neutrons, are grouped into a category known as baryons. Baryons are the large sub-atomic particles produced in fundamental nuclear reactions after the Big Bang.

There is a tendency for heavy elements to be associated with some MWT mineral deposits! Lead (in the form of galena) is the most obvious element to do this; however, barium (in the form of barite or barium sulfate) is also a heavy element, which for reasons not totally clear, as is the case with lead, can also occur in abundance.

Barite, as mentioned above, is named for its weight (baryon (G.)= heavy). Barite is a heavy mineral and the element barium is also a relatively heavy element. One is more accustomed to thinking of heavy-element mineral

The story of barytes. A 1917 booklet on barite and barite mining. Barite, in the early 20th century became a valuable commodity. With the expansion of the petroleum industry as a consequence of proliferation of the automobile, barite became indispensable in the process of deep drilling for oil. Barite prospecting and mining in the Ozarks became an important activity in the 1920s and this has continued until recently. A minor use for the mineral is also in the form of "barium cocktails" administered in the taking of gastrointestinal X-rays, and barite also serves as a substitute for lead oxide in the formulation of paint. Deposits in Missouri's Washington County became one the world's major supplier of high-quality barite. In the 1920s, barite was prospected for extensively over the Ozarks, and numerous deposits, small and large, were mined from the 1920s to the '50s, some of which yielded distinctive and attractive mineral specimens.

occurrences associated with hydrothermal or hyperthermal environments. Occurrences of heavy elements like silver, gold, and mercury, as well as the rare earth elements are usually found associated with these high-temperature, deep-seated occurrences. Barium minerals likewise would be thought to occur in deep-seated, hydrothermal environments and they do sometimes, but barite more commonly is associated with sedimentary rocks, as in MVT deposits.

Hoisting barite by hand from a "diggings" in Washington County, Missouri. The man at the left is cobbing barite to remove rock and other impurities. At the right is a rocker box, used to remove red clay attached to the barite.

Hauling cleaned barite to market.

Barite deposit exposed in fractured bedrock (dolomite) in a road cut near Hillsboro, Missouri. Barite generally is associated with residual red clay which occurs just above bedrock. Here it occurs in bedrock in Lower Ordovician dolomite (or dolostone).

Closer view of white barite in dolomite. The grey mineral is a mix of barite and calcite.

Closer-yet view of above barite and calcite masses.

Northeastern Ozark, Washington County, Barite Occurrences

All of these barite (or barytes) specimens came from Washington County, Missouri, one of the major barite producing areas of the world, although little mining activity currently is being done there. The Washington County barite district covers a large portion (most) of Missouri's Washington County, some 2,500 square miles. Galena was first mined in this district, starting in the early 18th century, when the area was under both France and Spain. Mining was done by digging into the red clay until reaching bedrock, which occurred ten to fifty feet below the surface. Galena would be found in the red clay occurring just above bedrock or sometimes concentrated between rock pinnacles, the edge of which sometimes produced the richest deposits. Barite occurs extensively with galena; however, at the time, it was worthless. The barite was of interest, as its presence indicated the likelihood of galena being present and large masses of barite often contained galena crystals in their interior. Not until the early part of the 20th century did barite become a useful commodity when numerous uses were found for it, including its use as an additive to paint and rubber and in the petroleum industry as a material found indispensable in the drilling of deep holes in the process of oil exploration.

The Potosi (Mine Au Breton) mining area around 1914. Mine Au Breton was one of the earliest regions to be developed as a mining region in what would become the United States. Mining for lead (galena) began in this area around 1720 and continued until around 1980, but in the 20th century, barite became more important than lead in the district. The mining areas are those showing areas of red clay. This red clay (tallow clay), is widespread in Washington County and is the source of most of the barite. (Potosi, Missouri, is now the county seat of Washington County).

Barite pit, Old Mines,1960. Barite was mined extensively by mechanical methods in the Washington County area from the late 1920s until the 1980s. This is one of the mined areas with red clay and many rocks.

Part of a large barite mining area near Tiff, Missouri, 1960. Tiff is a local term in Missouri for barite.

This is the same area as shown in the previous photos in 2012. It now supports a variety of wildlife, which includes some wet areas with absolutely lovely green bull frogs.

Stromatolite-chert. This peculiar chert boulder has this configuration as a consequence of it being a mass of silicified stromatolites. Stromatolites in some manner appear to have aided in the deposition of barite and in the mineralization process. This is looking at a cross-section of what are known as digitate stromatolites.

Diagram of digitate stromatolite reef. The holes in the rock shown in the previous photo were the site of the purplish "fingers" shown in this diagram.

Bladed barite,Washington County.

Windlass being used to mine barite in the foreground; in the background is a mined area with the ever-present red clay (tallow clay). Washington County, barite region, 1914.

Large barite mass, Old Mines, Missouri,

Right, barite from Palmer Missouri, left-from Old Mines Missouri. The Palmer barite area is in the western part of Washington County. Barite crystals found here are usually not very attractive.

Use of a hand windlass in mining. By digging down into the earth and then digging laterally into the red clay, chert, barite, and galena were brought up by the windlass. By use of this labor-intensive technique, large areas could be mined for both galena and later for barite. Over many parts of Missouri's Washington County, numerous depressions formed from the collapse of small, hand-dug mines dating from the late 1700s to the early 1900s can be seen.

Clear blades of barite with some of the ever-present iron-oxide-rich clay stain. (Value range F.)

These barite crystals from near Tiff, Missouri, occupy a quartz druze. Tiff is a local Missouri name for barite (or barytes), but in other mining areas, it is sometimes a name for calcite. Originally, tiff referred to any worthless mineral encountered during the mining of lead.

Additional small, clear barite crystals from dolomite vugs.

Small, narrow, clear barite crystals formed in cavities in dolomite (vugs). (Value range G, single crystal.)

White barite in a druzy cavity. (Value range F.)

An especially nice group of barite crystals on a limonite slab. (Value range F.)

Translucent barite crystals occupying a cavity.

Group of bladed barite crystals. Clusters like this in the Washington County barite district are found in red clay, which occurs just above bedrock.

Blades of barite crystals containing orange-ish iron oxide as an impurity. (Value range F.)

Glass tiff, Shirley, Missouri. Clear barite crystals from Missouri are known by the miner's term of "glass tiff." This cluster came from barite diggings near Shirley, Missouri, in western Washington County. Glass tiff is uncommon in most of the Washington County deposits, most of it having come from an area just north of Lake of the Ozarks. *Courtesy of St. Louis Science Center.*

Sphalerite crystals on barite, Potosi, Missouri. Sphalerite is normally not found associated with Washington County barite—an exception is in the Potosi area, where crystalline dolomite of the Eminence Formation contains sphalerite.

Small barite crystal group (crested barite) on massive barite. Old Mines area, Washington County, Missouri, barite district.

A Gallery of Ozark Barite Specimens

These barite crystals came from various portions of the Ozark Uplift of Missouri other than the Washington County barite district. Each of these different barite occurrences yields crystals distinctive in their appearance—distinctive enough so that a knowledgeable collector can often recognize where they originated. Although these specimens are all crystals of barite, each district's crystals can have their own distinctive "signature," a property of the mineral kingdom known as habit. The phenomena of mineral habit is not totally understood; it may have to do with the presence (or absence) of trace elements or be related to the rate at which the crystal formed. Mineral habit, however, enables a knowledgeable person to locate where in a particular region, like the Ozarks, a particular crystal came from.

Another habit of barite crystal from the Lake of the Ozarks area collected in the 1940s. *Joe Schraut collection.*

Grey barite with white tips characterize barite crystals from the Proctor Creek diggings located near Lake of the Ozarks. This area was extensively prospected for barite in the 1920s and produced some distinctive barite specimens.

Large specimen from the Proctor Creek diggings collected in the 1940s. *Joe Schraut Collection.*

A distinctive habit of barite from Texas County barite deposits. Large specimens from this barite region are rare, most are thumbnail size. (Value range E.)

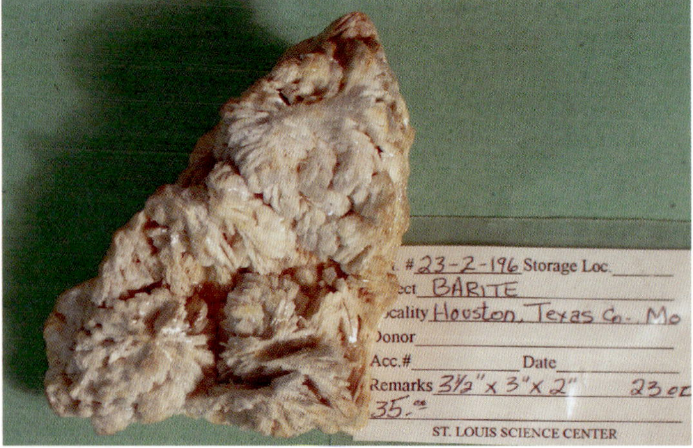

White barite, Texas County, Missouri. A barite occurrence was mined in the 1930s in Missouri's Texas County, which produced numerous small, sky-blue crystals and larger white ones like this. Often, the blue crystals are tipped with white barite. Crystals from this locality, as well as those form other small mines and prospects, are often distinct and differ from more commonly seen specimens like those from Missouri's Washington County. The presence of these distinctive minerals is known as crystal habit. They also can be more desirable as they are rarer. (Value range E.) *Courtesy of the St. Louis Science Center.*

Distinctive habit of bladed barite from Pioneer Mine, Jefferson County (?), Missouri. *Courtesy of St. Louis Science Center.*

Cleavage fragment broken from another blue barite mass. Franklin County, Missouri. Currently (2014), sky blue barite crystals similar to this are coming from Morocco.

Blue barite (cleaved pieces). Masses of sky-blue barite occur along the middle portion of the Meramec River, a spring-fed Ozark stream that enters the Mississippi just below St. Louis. These barite masses are unattractive on the surface, but when cleaved, reveal a single crystal of sky blue, often clear barite. The author has been told that blue barite like this, mined at Anaconda, Franklin County, Missouri, west of St. Clair Missouri, became the basis of "Woodcock's Ozark Rock Curios," a prominent attraction on Route 66 west of St. Clair. (Value range F, single specimen.)

Texas County barite specimen, different lighting.

Thumbnail specimen of blue, Texas County barite. Large specimens like that in the previous photo were available when the barite was being mined. Today, one can consider themselves lucky if a small specimen like this is found. (Value range F.)

A single crystal of blue barite showing cleavage faces, Franklin County barite district. Small galena occurrences occur over the middle part of the Meramec River and these were discovered and worked (for lead) in the early and mid-19th century. The blue barite that accompanies galena in this district at the time was considered worthless. Mineral specimens from old mining districts like this are often of greater value than are specimens from more modern mining operations. Usually, few specimens were saved and they also have historic as well as mineralogic significance. (Value range E.) *Joseph Schraut collection.*

Barite crystals on the fossil of an early cephalopod (ellesmeroid) from the same area as the previous photo. Barite crystals formed in a cavity in the living chamber of this early relative to the nautilus, squid, and octopus. This combination of mineral specimen and fossil is a rare one. From the Meramec, Franklin County lead-barite district. (Value range E.)

Glass Tiff. Clear barite crystals have been given the name of glass tiff by barite miners. Tiff is an Ozark term for barite. These clear barite crystals come mainly from the northern Ozarks barite district just north of Lake of the Ozarks. (Value range E.)

Circle Mines barite. Circle's refer to circular solution structures developed in limestone and dolomite. Sometimes these can contain significant mineralization. White aggregates of bladed barite crystals occur in a mineralized area where the mineralization occurs in circles (hence the name, circle mines). The crystals occur in dolomite and not in red clay as is the case with many other barite occurrences in Missouri and elsewhere. Small azurite and malachite spheres can also occur between the blades; the blue azurite spheres, when exposed to the air, however, alter to green malachite. (Value range G.)

White bladed barite (right, with small malachite crystals) and barite cleavage surface (left). Circle Mines, Cole Co., Missouri (Value range G.)

White, bladed barite with small pyrite pseudomorphs. What originally were small pyrite crystals have oxidized to iron oxide (limonite.) Circle Mines, Cole County, Missouri. (Value range G.)

Yellow barite crystal, Rolla, Missouri. Clusters of clear to translucent barite crystals like this occasionally are found in quarries and road cuts in the central Ozarks. These occurrences appear to be stratigraphically controlled in that they are generally found associated with Lower Ordovician strata. (Value range F.)

Close-up of specimen at the right in the previous photo. Buckshot Mine, Morgan County, Missouri. (Value range F.)

Non-MVT Barite

Barite can be a ubiquitous mineral; here are a few Ozark occurrences in odd geologic settings.

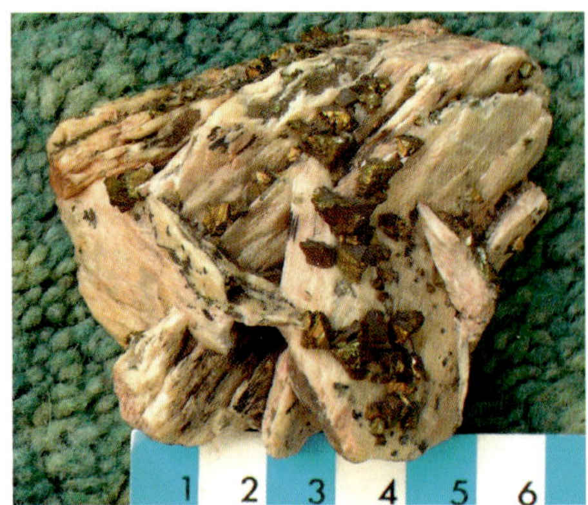

Bladed barite covered with chalcopyrite crystals and associated with Precambrian hydrothermal iron deposits. Pea Ridge Iron Mine, Missouri. Bladed barite, almost identical to that found in shallow surface occurrences in red clay, occur in this deep mine associated with massive deposits of hematite and magnetite at 3,000 feet below the surface.

Boulders of weathered Pennsylvanian age limestone, North St. Louis County, Missouri, an occurrence of the type that produced the crystals shown on page 75. A thorough examination of such transient outcrops may yield minerals or fossils. Earlier collectors not only had really sharp eyes, but also combed most available outcrops, a feat more difficult to do in the present-day U.S.

White barite, Rosiclare, Illinois, fluorite district.

Orange barite, St. Louis County, Missouri. (Value range F.)

Orange barite, St. Louis County, Missouri. Orange barite like this occurred in Pennsylvanian age limestone and was found in the 1940s in the northern part of the St. Louis area. Like many odd and isolated mineral occurrences, specimens like this are now seen only in old mineral collections.

Well organized and documented mineral, fossil, and rock collections should be considered what they are, that is attractive documents of past mining activity, as well as documents of the earth's geologic past.

Barite in geologic environments outside of the Ozarks.

Pink barite, Cedar Creek, Callaway County, Missouri. A variety of habits of barite occur throughout the Ozarks. This is a portion of a barite mass found inside a chert boulder in an area not known to produce barite. It resembles pink feldspar (orthoclase)

Pink barite cleavage fragments, Cedar Creek, Missouri.

Clear barite, Linwood Mine, Buffalo, Scott County, Iowa. This clear barite from limestone beds in Iowa came onto the mineral market in 2013 in quantities. (Value range F.)

Barite roses. These barite crystals occur in red sandstone of Permian age, many of which are found in the vicinity of Oklahoma City, Oklahoma. Only in the sense that they occur in sedimentary rock can these barite roses be considered a type of MVT mineral.

Group of barite roses, Oklahoma City, Oklahoma.

Gemmy blue barite crystals. Barite crystals like these occur in large, iron-rich septarian concretions that weather from late Mesozoic (Cretaceous) shale beds of the high plains of both the U.S. and Canada. These crystals often are associated with honey-yellow calcite–they can be quite attractive, as well as desirable. (Value range F.)

Barite crystals can occur in a variety of sedimentary as well as igneous rock environments. One of the best known of these are barite "roses," found locally in abundance in sandstone of Permian age in Oklahoma. Another occurrence of collectable barite crystals are found in concretions, which occur in shales of late Mesozoic age (Upper Cretaceous) found in the high plains of Colorado. These are also found in the Dakotas, as well as in the intermountain basins of Wyoming, western Colorado, and Montana. These barite crystals are usually clear, gemmy, and therefore highly collectable.

Septarian concretion sphere. This is a sphere made from a concretion of the type, which produces the above gemmy barite crystals. The barite crystals occur associated with the yellow calcite in interior cavities of septarian concretions. Most septarian concretions of the west, however, lack such barite crystals. (Value range F.)

Barite of the Southern Appalachians

Bladed barite occurs associated with red, residual clay in the southern Appalachians of Georgia and Alabama. It occurs here in a manner similar to that of the Ozarks of Missouri. One of the more unusual aspects of these occurrences is the presence of barite replaced fossils—especially the puzzling archeocyathids found near Cartersville, Georgia.

Cluster of light blue barite crystals, Cartersville, Georgia.

Blue barite, Cartersville, Georgia. A similar crystal of light blue barite. (Value range G.)

Blue barite, Cartersville, Georgia. This barite, like occurrences of blue barite in the Missouri Ozarks, is associated with thick beds of residual clay. Cartersville barite can occur as well-formed, clear, light blue crystals; large ones are rare. (Value range G.)

Archeocyathid preserved in barite. Archeocyathids represent a group of extinct life forms (possibly even an extinct phylum) which resembles both sponges and/or corals but is nether. Archeocyathids are unique to the Cambrian period of geologic time, which was that part of geologic time when multi-celled animals and plants first became abundant in the fossil record. Barite occurrences near Cartersville, Georgia (northwestern part of the state), have yielded these barite replaced fossils. The setting of the Cartersville barite occurrences in the Appalachian Mountains, like the Austinville, Virginia, zinc occurrences (Chapter Five) is identical with some MVT mineral occurrences in the U.S. Midwest. The Cartersville barite occurrences are associated, like those in Missouri, with thick, red clay, formed from deep weathering of carbonate rocks. (The Shady Formation, which was deeply weathered to produce the clay, is of Lower Cambrian age.) (Value range E.)

Thick, dolomite beds of the Shady Formation eroded into pinnacles and exposed in a barite pit (Paga Pit) near Cartersville, Georgia.

Thick, red residual clay is the host material for barite occurrences at Cartersville, Georgia. This occurrence is very similar to barite occurrences in the Ozark Uplift of southern Missouri. The pinnacles are erosional remnants of dolomite. The dolomite is the Lower Cambrian Shady Dolomite.

Barite replaced archeocyathid, which shows the shape of the puzzling archeocyathid organism. Cartersville, Georgia barite district.

Cartersville barite district, 1965.

Archeocyathids. Another group of barite replaced archeocyathids from the Cartersville, Georgia locality. There is something about fossils found in mineral deposits, which, in an intangible way, speak about the great age of the fossils themselves. Here are two seemingly incongruent phenomena—that of fossils (and the life they represent with its seemingly fleeting impermanence) and crystals (of insoluble, inorganic compounds) representing permanence and vast time spans, occurring together.

Shady Dolomite exposed in barite pit (Paga Pit, Cartersville, Georgia).

Pinnacles of dolomite weathered from the Lower Cambrian Shady Formation, Cartersville barite district, 1965.

Bladed Barite from Morocco

Paleozoic limestone forms major portions of the Atlas Mountains of Morocco and locally can yield crystalline barite masses similar to those of the southern Appalachians and the Ozarks. Associated with this barite can also be galena, sphalerite, smithsonite, and vanadinite. The rocks in which these minerals occur have been involved in mountain building—tectonic activity that formed the mountains of northern Africa—but as with occurrences of these minerals in the U.S. Midwest, the associated rocks are limestone and dolostone.

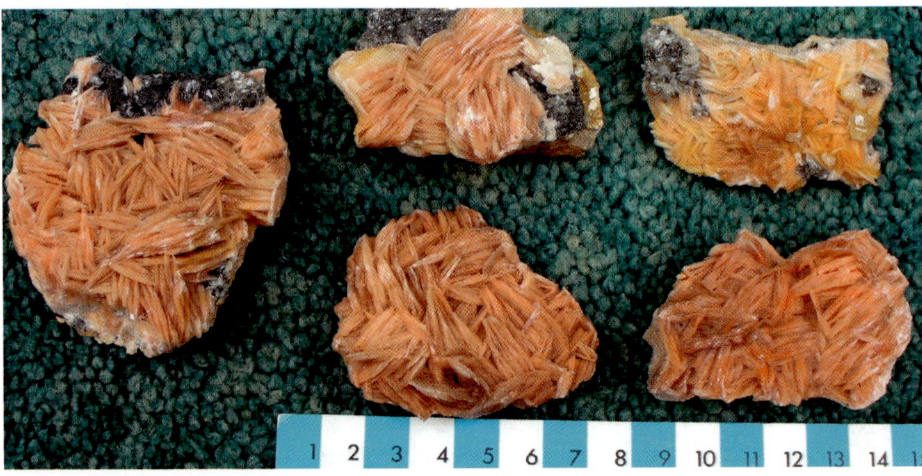

Group of bladed barite crystals from the Atlas Mountains of Morocco. Barite occurs here with weathered limestone but occurrences in a drier climate have produced some minerals that are absent from similar deposits in the U.S. Midwest.

Single cluster of bladed barite crystals. Mibladin, Morocco.

Bladed barite on which are perched cerrusite (left) and calcite (right) crystals. Mibladin, Morocco. (Value range G.)

Large slab of small barite crystals, Atlas Mountains, Morocco. (Value range F.)

Vanadinate on barite.

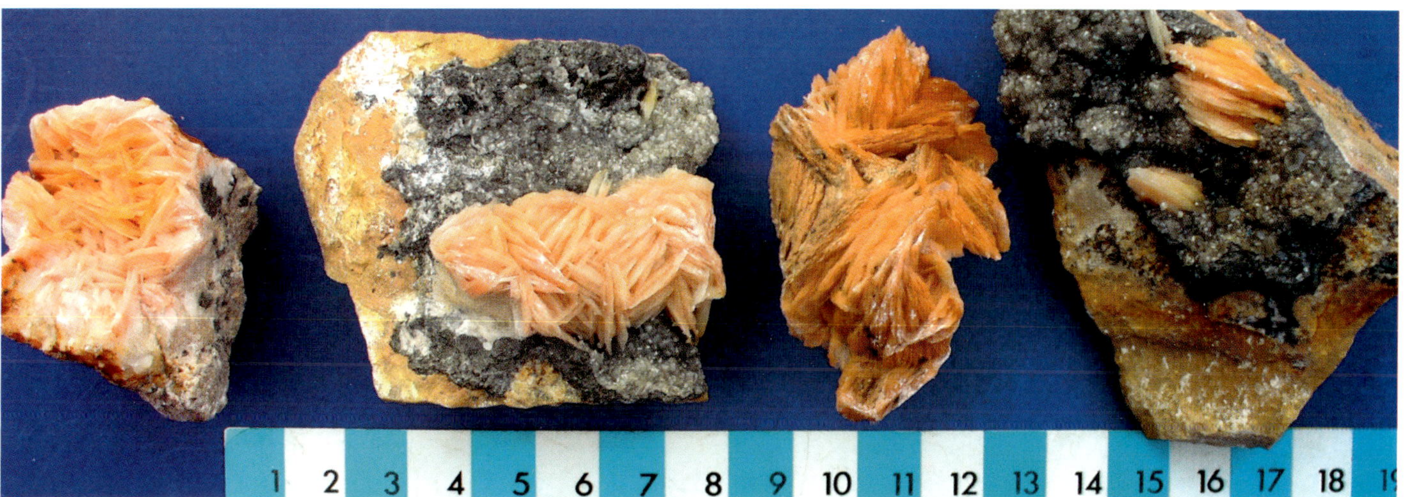

Barite and specimens of barite on cerussite, Mibladin, Morocco.

Vanadinite on barite. M'Fis Mine, Mibladin, Morocco. Vanadinite is a mineral not found associated with American MVT minerals. It appears associated with MVT minerals formed under a drier climate. (Value range F for single specimen.)

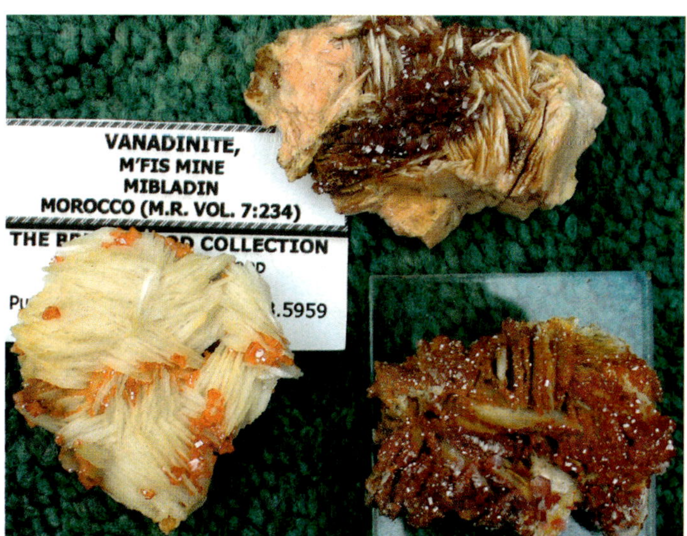

Group of vanadinate-barite specimens.

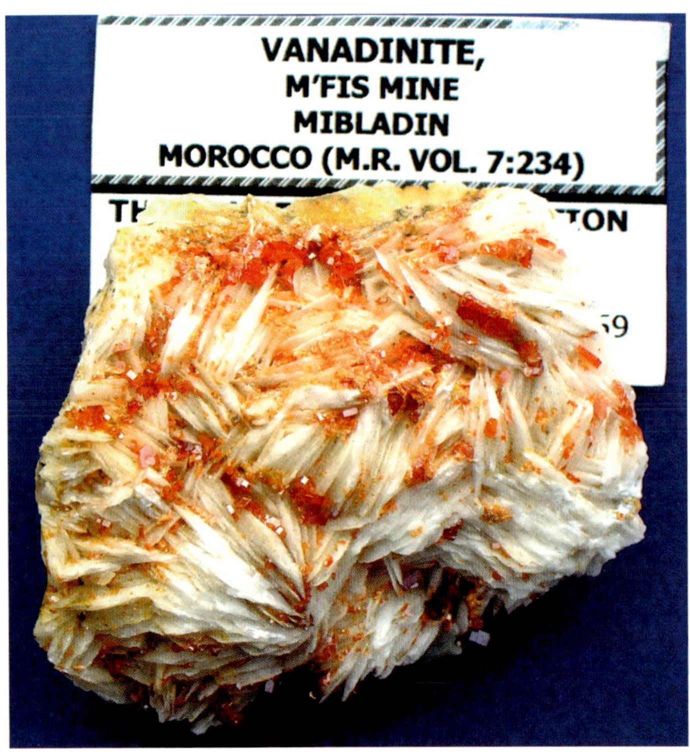

Vanadinate on barite. The attractive combination of these two minerals came from the M'Fis Mine, Mibladin Morocco—they came through the Tucson Show. (Value range F.)

Vanadinate on white barite covered with secondary grey hemimorphite. These minerals are associated with Mississippian limestone in the Anti-Atlas Mountains of Morocco; this bladed barite is similar to that which occurs associated with red clay and carbonate rock of the Ozark Uplift of Missouri.

Barite on cerussite and galena, Mibladin, Morocco.

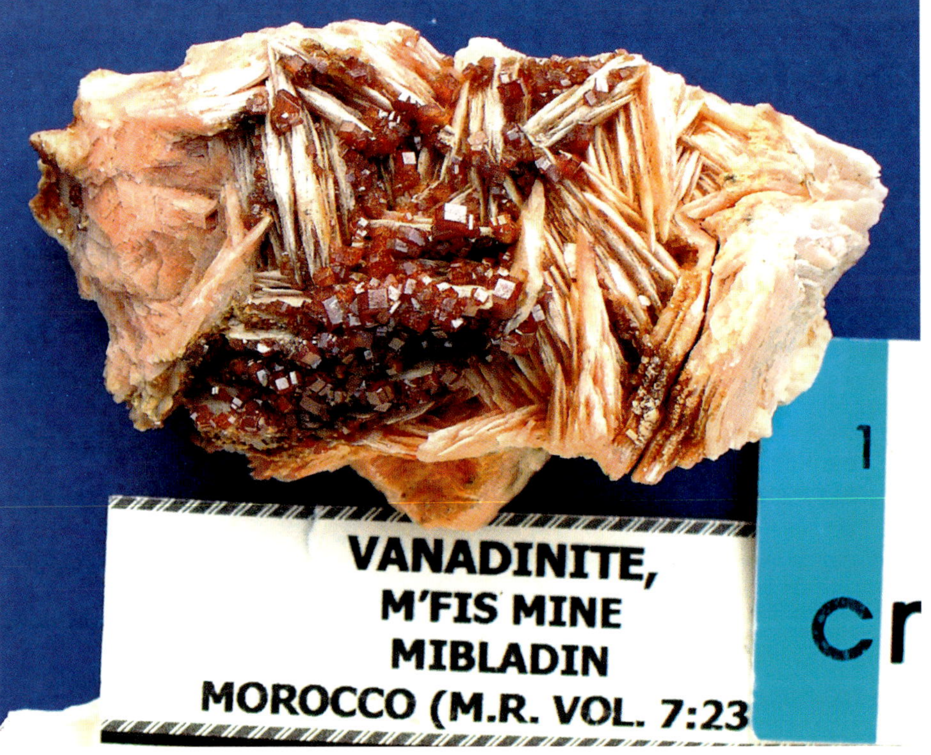

Vanadinite crystals clumped on barite. The association of vanadium minerals with bladed barite is a variant on MVT bladed barite occurrences. These gemmy crystals are not found with any known North American MVT occurrences, but otherwise the occurrence is quite similar.

Barite from Elsewhere!

"Elsewhere" can mean many parts of the world, as barite is a widely occurring mineral.

Greenish barite with malachite specks. Shaguluwe Mine, Katanga Province, Zaire, Africa. (Value range F.)

Blue barite, El Solar Mine, Taxaco, Mexico.

Barite on sphalerite, Peru. Barite occurs here in a well-documented hydrothermal or hyperthermal mineral occurrence. Veins of barite sometimes can be associated with such deposits. Unlike epithermal MVT barite occurrences, such veins usually continue with depth. Cerro de Pasco, Peru.

Close-up of above specimen. (Value range F.)

Yellow, gemmy barite on quartz. This beautiful gemmy barite came from China recently (2012) through the Tucson show. (Value range F.)

Witherite

Witherite is barium carbonate, a barium mineral that is more soluble and reactive than barite and sometimes is associated with it. Witherite occurs with barite in the southern Illinois Fluorite district, but otherwise is rare or non-existent in U.S. Midwest MVT deposits.

Witherite, barium carbonate. Rosiclare, Illinois fluorite district (Value range F.)

Celestite (Celestine)

Celestite is strontium sulfate. Strontium, a group II element on the periodic table (as is barium) lies just above barium on the table—hence it has similar chemical properties. Celestite might be expected to occur associated with barite, but it usually does not. Celestite does occur with barite in the southern Illinois (Rosiclare, Illinois, fluorite district), but like the barite with which it occurs, it is uncommon there. Celestite, often as fine crystals, occurs in cavities associated with Silurian dolomite beds of northern Ohio and southeastern Michigan in what is known as the Findlay Arch, an extension of the Cincinnati Arch. Quarries in these dolomite beds yield excellent light blue crystals of celestite, along with brown fluorite crystals (Chapter Eight).

Celestite, Maybee, southeastern Michigan, Findlay Arch. (Value range G.)

Celestite (CelestinE.) Strontium sulfate. This strontium mineral is similar to gypsum in that it is a sulfate. Strontium is also chemically similar to barium, both occurring in the same vertical row of the periodic table yet the two sulfate minerals are rarely found together.

Celestite, Stoneco Quarry, Northern Ohio. (Value range F.)

Celestite (Celestine) from Madagascar

Celestine similar to that of the Findlay Arch occurs associated with Jurassic limestone of Madagascar. Numerous specimens of this attractive mineral have entered the geo-collectable market from this island country through the Tucson show.

Celestite in a vug or geode. These egg-shaped specimens of Madagascar celestite came through the Tucson show in quantities in 2008.

Celestite, Majunga, Madagascar.

Fossils (Other than Archeocyathids) Associated with Barite Deposits

Peculiar molluscan fossils of Cambrian age occur with barite in the Ozark Uplift of Missouri.

Snail-like fossils (*Scaevogyra* sp.) These left-handed-coiling "snails" (paragastropods) are peculiar in a number of ways. They are associated with quartz druze and barite in the Washington County barite district.

Typical quartz druze associated with barite. Tiff, Missouri.

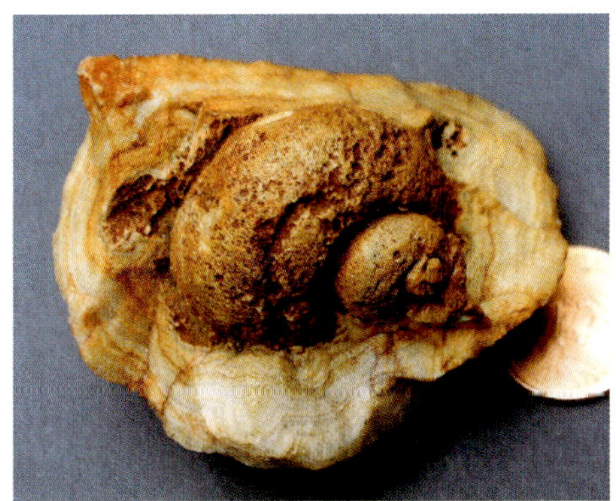

Flip side of the above druze showing specimen of the paragastropod *Scaevogyra* tucked inside the druze.

Scaevogyra sp. Specimen of this paragastropod next to a large druze. Ozark Trail, Black River, Missouri.

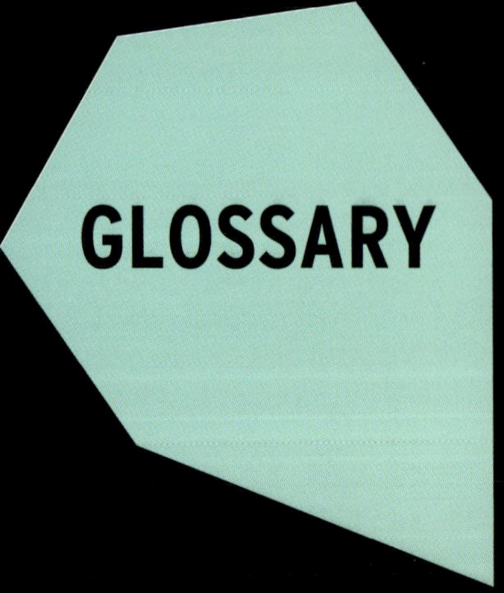

Barium

Barium is element number 56 on the periodic table, it is a heavy element, which are generally relatively rare. Heavy elements are rare because the mechanisms which produced them, extreme stellar explosions known as Supernovas are relatively rare events in a galaxy compared to Novas which are a more common phenomena. Calcium and Magnesium are in the same group on the Periodic Table (Group II) as barium but are much more common as they were produced in a Nova. As is the case with lead (element number 82), the abundance of barium in a non-hydrothermal MVT geologic environment in the Ozarks is a geochemical puzzle.

Density

Something that is heavy for its size as is the case with galena and barite. Density is weight per unit volume; something that is small but heavy has a high density.

Gangue mineral

A mineral that is associated with a mineral having commodity value, but which in itself has no value as a commodity.

Lithophile element

An element which is attracted to and associated with the common elements of the earth's crust like silicon, aluminum, sodium, calcium, and potassium.

Pegmatite

A coarsely crystalline mass containing (sometimes) unusual crystals. Pegmatites resemble coarsely crystalline granite and are found associated with granite masses.

Residual (red) clay

Iron oxide bearing clay that can be associated with barite deposits. Such clay formed as a residue left behind from the weathering of a larger mass of rock, which in the case of barite deposits often is limestone or dolostone.

Sulphophile element

An element that is chemically attracted to and associated with sulfur.

Vugs

Cavities or large holes in rock (usually limestone) or dolomite. Vugs can harbor crystals, including those of barite—sometimes crystals in the vugs can be especially nice and well formed and therefore collectable.

Chapter Four Resources

Clark, Allen W., 1917. *The Story of Barytes.* DeLore Baryta Company, St. Louis, Missouri

Cozzens, Arthur B., 1941. *Gopher-hole barite mining in Washington County, Missouri.* Illinois Academy of Sciences Transactions. Vol. 34, No. 2.

Kile, Daniel, 2008. *Barite-Colorado in American Mineral Treasures*, G. A. Staebler and W. E. Wilson, editors. Lithographie, LLC., East Hampton, Connecticut.

Tarr, William A., 1918. *The barite deposits of Missouri and the Geology of the Barite District.* University of Missouri Studies. Columbia, Missouri.

ZINC MINERALS

For reasons not understood, zinc minerals are one of the most common ones found in MVT deposits—not only in the Mississippi Valley region, but world wide where zinc minerals are associated with thick beds of limestone and dolomite, especially that of Paleozoic age. The most common zinc mineral is usually sphalerite (zinc sulfide), where it may be associated with galena and sometimes with barite. Calcite and dolomite crystals also are found commonly, but these might be expected, as they are frequently occurring minerals in all carbonate rocks.

Headframes of mines in the Tri-State (Missouri, Kansas, Oklahoma) mining region, once one of the world's largest producers of zinc. This MVT deposit had much of its mineralization as the mineral sphalerite. However, some of the mines produced smithsonite, a secondary zinc mineral formed from the oxidization of sphalerite. These mines include separate rail-loading facilities; however, many individual mines of the district were interconnected to form a vast network of underground workings. These mines, being below the water table, are all water-filled today.

Limestone beds, especially those of Mississippian age in the U.S. mid-continent often contain small masses or localized pockets of sphalerite, a mineral composed of zinc sulfide. Large deposits of this mineral can occur in limestone beds like these and have been the basis for extensive mining operations like those of the Tri-State region of Missouri, Oklahoma, and Kansas.

Sphalerite (zinc sulfide) is mentioned in geological exploration done prior to the mid-19th century in work like that done by David D. Owen, especially in Iowa. This outcrop of Mississippian age limestone occurs just below the Iowa line in northeastern Missouri; it contains a number of small pockets of zinc blende (another name for sphalerite).

Sphalerite in a calcite geode. Geodes are peculiar, often hollow rocks containing crystals in their interior. Sometimes they can contain sphalerite (zinc sulfide), especially geodes from northeastern Missouri and southeastern Iowa. Sphalerite is the dark crystal. Localized occurrences of MVT minerals like this, often exhibit a habit that is different from major occurrences like the Joplin Tri-State region and thus are more desirable to some collectors. Other collectors, on the other hand, only want minerals from well-documented mining areas—whatever floats your boat! (Value range F.)

Smithsonite, Granby, Missouri. Smithsonite is zinc carbonate, a secondary mineral formed from the weathering and oxidization of sphalerite. Smithsonite was the second-most important zinc mineral mined in the Tri-State region.

Thumbnail mount of black sphalerite from near Arnold, Missouri, just south of the St. Louis area. Thumbnail mounts enable a person to have a large and diverse mineral collection without too much expense or taking up too much room—there is a lot to be said for them.

(Right) Single sphalerite crystal from just south of St. Louis, Missouri. (Left) Hannibal Missouri, all from Mississippian age limestones.

Dark (blackjack) sphalerite (left) in a calcite geode.

Blackjack sphalerite. Blackjack sphalerite is sphalerite of a dark hue, usually either very dark brown or black. The presence of iron in the crystalline lattice is the element producing the black color. Pure zinc sulfide is colorless.

Vugs are cavities that occur in rock, often this rock being limestone. Sphalerite-filled vugs like this are found sporadically in limestone of Mississippian age where they may grade into geodes (which appear to be related to vugs). These came from various places in the northeastern Ozarks, southwest of St. Louis, Missouri, and (left) Hannibal, Missouri quarry. (Value range G, single specimen.).

Sphalerite from northeastern Missouri (Hannibal quarry) where it occurs in vugs.

Tri-State (Missouri) Mining District

Zinc minerals were known to occur in the Mississippi Valley Region since mineral exploration began in the 18th century; mineral exploration starting in this region with the 16th and early 17th century search for gold and silver. Sphalerite was noted in its association with galena in the upper Mississippi Valley region in the 19th century, where the three states of Iowa, Wisconsin, and Illinois come together in what became known as the upper Mississippi Tri-State mineralized region. Sphalerite was also a frequently

Head frame of a zinc mine near Granby, Joplin Tri-State area, 1899. In the late 19th century, extensive deposits of zinc minerals were discovered in southwestern Missouri. By the beginning of the 20th century, mining of both primary sulfide zinc minerals (mainly sphalerite) and galena was being done on a large scale. Some of the minerals found near Granby were colorful secondary zinc minerals, like smithsonite.

Oronogo Circle, Oronogo, Missouri. The Oronogo (*Ore-or-no-go*) mines were developed in the late 19th century and mined both primary sulfides and secondary zinc and lead minerals, like smithsonite and anglesite. Like the Granby mines, the secondary minerals from Oronogo Mines were somewhat unique for what would become the Missouri Tri-State mining region. (The other Tri-State mining region is that of Illinois, Wisconsin, and Iowa.) In the late 19th century, mineralization west of Missouri was poorly known. In the early 20th century, extensive mineralization was discovered to extend westward into southeastern Kansas and into what, at the time, was known as Indian Territory (became Oklahoma in 1907). Activity in this part of the district didn't take place until after the Great War (WWI) when the more deeply buried mineralized zones were mined.

Chat (crushed limestone and chert) pile from zinc and lead mining in southwestern Missouri has been used extensively for making roads and also for rock along railroad tracks (ballast). Many persons note the calcite "pebbles" found in this material by their cleavage surfaces, which sparkle in the sun along gravel roads and railroad tracks–not only in southwestern Missouri but over a large portion of the U.S. Midwest.

U.S. Highway 66 traversed through the Joplin-Tri-State mining region in the 1920s through the 1960s. The chat piles along Highway 66 were once a landmark as one traversed through southwest Missouri. "Get your kicks on Route 66–now you go thru St. Louie, Joplin, Missouri(ee)–Oklahoma City is mighty pretty" so the song goes.

occurring mineral in the Pierson Creek lead mines of the early 19th century, south of the present city of Springfield, Missouri. Here, dark brown sphalerite occurs as well as gemmy red crystals on slabs of limestone. At the time, sphalerite had no use, zinc itself being a recently discovered element of which not too much was yet known. As the 19th century progressed, zinc minerals became more and more available on a large scale and were found to be especially abundant in mines worked for galena in both Arkansas and southwestern Missouri, southeast Kansas, and adjacent Indian Territory (Oklahoma). This Tri-State mining district closed in the early 1980s. Many of the mine dumps have been reclaimed and the mines, most being below the water table, are now filled with water. The numerous primary minerals scattered in collections worldwide and pictures preserve its legacy. A legacy that got some negative publicity with the 2011 Joplin tornado was a tragedy that destroyed much of the town and killed over 160 persons. The tornado is claimed to have scattered lead-bearing mine tailings over parts of the devastated town and hindered rebuilding.

Small mines in the Joplin, Missouri area, 1900.

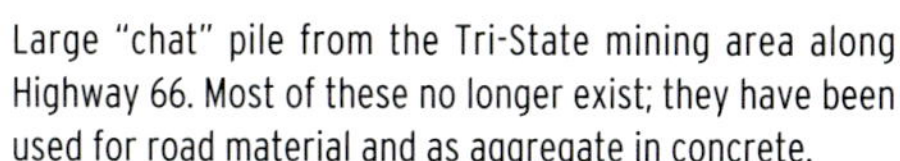

Large "chat" pile from the Tri-State mining area along Highway 66. Most of these no longer exist; they have been used for road material and as aggregate in concrete.

Chat piles and mine dumps in the Joplin district, c. 1955.

Zinc Minerals

As galena was usually associated with the zinc minerals, lead was the valuable commodity, but more attention began to be taken toward the (usually more common) zinc minerals, a gallery of which is shown here. By the late 19th century, a number of significant uses were found for zinc, including its use as a coating of iron to prevent rusting (galvanizing), its use in the lining of tin cans, and its use in electrical batteries, as well as an interesting minor use in the making of early disk phonograph records.

Sphalerite with relatively rare barite crystals from the Tri-State, Joplin district. Barite is relatively rare in the Tri-State district, therefore, a specimen like this is desirable. (Value range F.)

Crystals of the two major minerals from the Tri-State district. Galena cubes are attached to sphalerite crystals. (Value range F.)

Typical Tri-State sphalerite crystals covering chert. Chert is especially characteristic of minerals originating from the Tri-State region. The rock in which the mines were located was exceedingly cherty and mineralization often was associated with chert breccias.

Average Tri-State sphalerite specimen with the usual "dings" from mining. Note the chert at the bottom left.

Chert slab covered with sphalerite crystals (Value range F.)

Typical Tri-State sphalerite on chert. (Value range F.)

Sphalerite on chert with octahedral galena crystals (Value range F.)

Somewhat dinged sphalerite crystals covered (top) with dolomite crystals. This is an average Tri-State specimen still readily available as of the time of writing, (2012). (Value range G.)

Dickite covering calcite. Here is dickite, which is less homogeneous and therefore less desirable. (Value range G.)

Dolomite crystals with blackjack sphalerite. (Value range F.)

Dickite covering calcite covering sphalerite. Dickite is a relatively rare mineral, especially when it covers calcite so evenly as this. (Value range F.)

Ruby jack (red sphalerite) on galena. Red sphalerite has less iron in it than does darker sphalerite (Black jack) and also is more attractive (to most collectors). Ruby jack is rarer than the other forms and usually occurs as small, gemmy crystals. Here the small crystals are perched upon yellowish colored sphalerite covering cubes of galena (Value range F.)

Light pink dolomite crystals on which are perched small tetrahedral crystals of chalcopyrite. This is a common association of Tri-State minerals and the light pink dolomite coupled with the crystals being on chert identifies the minerals as being from this region. (Value range F.)

Large sphalerite "Clump" on a slab covered with dolomite and chalcopyrite crystals. *Boente Collection. Courtesy of St. Louis Science Center.*

Numerous ruby jack crystals on chert. (Value range F.) *Courtesy of Stan Perry.*

Tri-State Fossils

A few spectacular fossils were found in the mines when they were working; especially noteworthy are the melon-like primitive echinoids (sea urchins) found near Webb City, Missouri.

Fossil primitive echinoid from Webb City (Tri-State) mine. Numerous specimens of this large (5 inches in diameter) early sea urchin were found in 1901 in chert beds associated with sphalerite mining. Of the genus *Oligoporous* sp., the surface of this echinoid is covered with a fine druze of sphalerite crystals. (Rare.)

Oligoporous sp. A small specimen of this echinoid preserved in chert from the northeastern Ozarks. *Courtesy of Don McKinnis.*

Washington County, Missouri Sphalerite

In most of the southeastern Missouri MVT deposits worked for lead; the mineral sought was either galena or anglesite. The earliest mining of galena took place at Mine LaMotte and Mine au Breton, both areas being worked as early as the 1720s. Near Mine au Breton (now Potosi, Missouri), sphalerite sometimes occurs as small gemmy crystals as well as a more granular form filling between dolomite clasts in a dolomite-sphalerite breccia, but otherwise zinc minerals were absent from these early-mined areas.

Sphalerite on pyrite from a vug. Sphalerite crystals from the Potosi area are relatively uncommon in MVT collections. These occur in vugs in Cambrian Eminence Formation dolostone. (Value range F.)

Vug with sphalerite (dark brown), pyrite (greenish), calcite (yellow), ferroan dolomite (red) and barite (white) from the Potosi, Missouri area. (Value range F.)

Sphalerite crystals on barite. Potosi, Missouri. (Value range F.)

Sphalerite and red dolomite in vugs.

Small, Localized Zinc Deposits

Prospecting for minerals in the Ozark Region was extensive in the late 19th and early 20th centuries. Some of the prospects led to mines, usually small ones that produced galena, sphalerite, copper, or barite (in the 20th century). A few of these produced minerals of a distinctive habit—others with only unattractive ore (from a mineral collector's perspective) with few or no crystals. Specimens from some of the small mines and prospects have entered collections and a few of these occurrences did produce exceptional specimens of a distinctive habit.

Sphalerite and calcite in vug in weathered dolostone. Alice Mine, Butler County, Missouri.

Sphalerite on quartz. Alice Mine, Butler County, Missouri.

Alice Mine, Butler County, Missouri, 1960. Many diggings like this worked small deposits of sphalerite, barite, and galena from the 1890s to the 1960s; they were widely scattered over the Ozarks. Sphalerite specimens from this locality left something to be desired.

Sphalerite in a Pennsylvanian age ironstone concretion. Grand River, Missouri. Sphalerite can occur with some frequency in ironstone concretions that weather from shale beds of Pennsylvanian age. Sphalerite was mined from Pennsylvanian age sink hole deposits of the northern Ozarks in the 1920s.
(Value range F.)

Secondary (Supergene) Zinc Minerals

Some of the first ores mined in what would become the large Tri-State mining region (Missouri-Oklahoma-Kansas) were in the form of what are referred to as supergene or secondary minerals—minerals formed from the weathering of primary sulfide minerals like lead and sphalerite. These secondary minerals were anglesite (lead sulfate), cerussite (lead carbonate), hemimorphite (zinc silicate), and smithsonite (zinc carbonate) They occurred near the surface and could readily be mined, as they were easily dug. It also

Hemimorphite is a zinc silicate. Clusters like this were associated in quantity with some of the earliest large-scale mining in the Tri-State district during the late 19th and early 20th centuries. (Value range F.)

Smithsonite, Oronogo Circle Mine. A mineral composed of zinc carbonate, smithsonite is formed from the weathering of primary sphalerite, which occurs in limestone and dolostone in MVT deposits. It's an example of what is known as a secondary mineral, an oxidized mineral formed from a more reduced one like sphalerite. (Value range E, from a classic locality.)

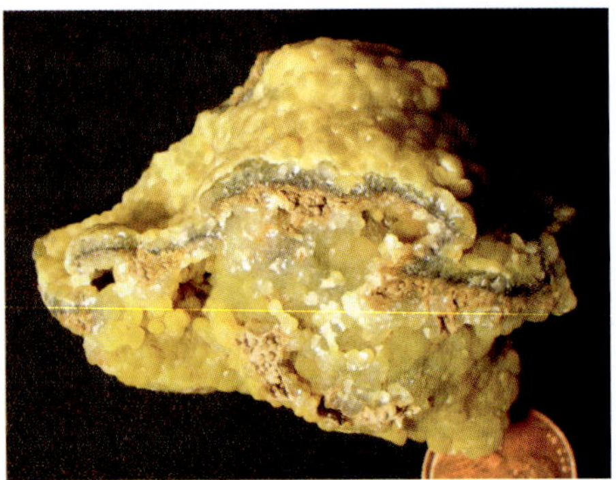

Botyroidal smithsonite on dolomite, Oronogo Circle Mines, SW Missouri. (Value range E.)

Granby, Missouri, smithsonite. All of these specimens from Granby, Missouri, came from old collections–they are quite collectable today. (Value range F.)

Limonite, part of a mass of ocher, which forms a gossan or "iron hat" capping a deposit of secondary minerals, like those of Granby, Missouri.

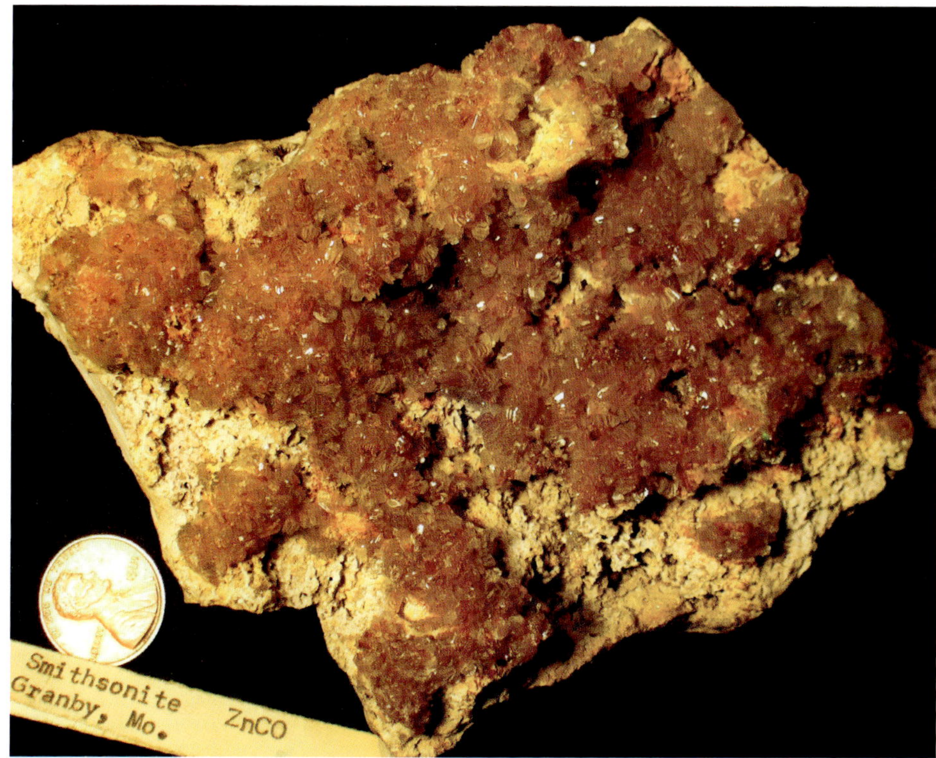

Smithsonite from Granby, Missouri. Attractive specimens of yellow smithsonite came from secondary minerals formed above the water table from the weathering of primary sulfides. These were mined extensively during the last third of the 19th century and early in the 20th. (Value range E.)

was relatively easy to extract lead and zinc from them compared with the more difficult process of smelting sphalerite and galena.

Sphalerite is zinc sulfide. In this form, it is stable if not exposed to air and water for a long period of time—that is, stable when it is un-weathered in the ground and incorporated in limestone or dolomite. When exposed to weathering near the surface, however, these minerals react with both oxygen and water, forming the secondary minerals mentioned above. Commonly associated with these secondary minerals are the weathering products of pyrite and marcasite. The top part of an ore body often has a concentration of limonite (or ocher) formed from the weathering of iron sulfides. Such a "cap" is also known as gossan or an "iron hat." These secondary minerals were some of the first minerals to be mined in the Tri-State region, especially near the town of Granby and Oronogo (*ore-or-no-go*) near Joplin, where attractive specimens occurred, including some of those shown here.

Rush, Arkansas A Significant Ozark Zinc Deposit Consisting Almost Entirely of Smithsonite

One of the largest zinc mining regions in the Ozarks was near the town of Rush, in Marion County, Arkansas. Here extensive deposits of honey-colored smithsonite, known as "turkey fat" ore, were extracted from Lower Ordovician Dolomite in bluff-side mines. In the late 19th century the Rush, Arkansas zinc district was a major producer of zinc. Associated with the deposits was cadmium, the source of the yellow color in turkey fat ore. The area is now within the boundaries of the Buffalo River National Scenic River and prospecting or collecting is prohibited.

Buffalo River, Arkansas. Incised into Lower Paleozoic strata, the St. Peter Sandstone forms high bluffs along this scenic river. Strata which yielded the zinc minerals at Rush, Arkansas, lie below this strata. This photo is upstream from Rush Creek, where the Buffalo River has not yet cut into the older strata.

Buffalo National Scenic River, Arkansas. Secondary zinc minerals like smithsonite and hemimorphite were discovered and mined in the hills along this river from the 1890s to the 1950s.

Like many streams that flow off the southern portion of the Ozark Uplift, the Buffalo River has a relatively steep gradient, sometimes with small waterfalls like this.

Rush, Arkansas zinc district, 1900. Outcrops of dolostone (or dolomite) are exposed along Rush Creek, a stream which flows into the Buffalo River of northern Arkansas. The Rush, Arkansas mines, early in the 20th century, were an important source of zinc mostly as secondary zinc minerals, especially smithsonite.

Same specimen as in previous photo. The yellow color (which comes from cadmium) is probably the most frequently seen form of this mineral, especially in the Rush, Arkansas, occurrence of northern Arkansas. (Value range F.)

Rush Creek (Morning Star Mine). The smithsonite was (is) associated with breccias formed in Lower Ordovician (Cotter) dolostone. One of the many mines along Rush Creek can be seen in the bluffs behind the large tree nearest to the man on the right. Mineralized breccias were found to occur at numerous places along Rush Creek which now is within the boundaries of the Buffalo River National Scenic Riverway administered by the U.S. Park Service.

Breccia associated with secondary zinc mineralization. MVT secondary minerals are generally associated with solution breccias like this. These made up most of what actually was mined as ore. The beautiful botryoidal or reniform surface of smithsonite and hemimorphite developed within cavities associated with such breccias.

Ledges of sandstone like this can form waterfalls or rapids where the river encounters them, especially those rivers which flow off of the southern part of the Ozark Uplift. Dolomite (or dolostone) beds, which occur between the sandstone layers, can be host for MVT mineralization, especially in the southern part of the Ozark Uplift.

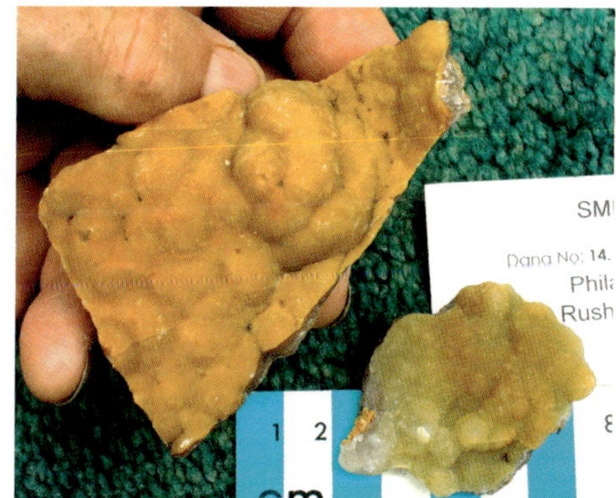

Turkey fat smithsonite (cadmium smithsonite) This yellow smithsonite gets its color from cadmium, an element often associated with zinc minerals.

Colorless smithsonite, Rush, Arkansas.

White or colorless smithsonite from the Rush, Arkansas, district. (Value range F.)

Same specimen as above, different view. (Value range F.)

Colorless smithsonite, Rush, Arkansas.

Colorless smithsonite.

Small spheres of yellow smithsonite. Morning Star Mine, Rush, Arkansas.

Blue-grey smithsonite with a well-developed botryoidal surface, Rush, Arkansas.

"Turkey fat" smithsonite with well-developed botryoidal or reniform surface.

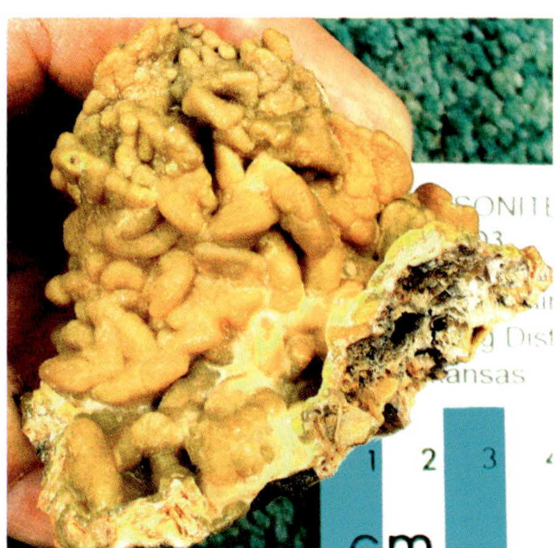

Some of the Rush, Arkansas, smithsonite minerals have a peculiar botryoidal surface like this. There is something about these specimens (like this one) that is somewhat disturbing, not only because they resemble a blob of fat but also because they also suggest a mass of yellow maggots. (Value range F.)

Smithsonite mass, Rush, Arkansas. The linear features seen here (smithsonite and hemimorphite-filled cracks) are typical of secondary minerals like those mined at Rush, Arkansas.

Grey smithsonite masses at Rush, Arkansas.

Secondary Minerals from Other Locations

Other than smithsonite from Granby, Missouri, and Rush, Arkansas, secondary minerals from the Ozark Region, as well as other parts of the Midwest, are not frequently seen, although they do occur. This is in part because these minerals were mined in the 19th century or early in the 20th, when there was less interest in minerals as collectables, but also because specimens usually were small. In other parts of North America, like in New Mexico and Mexico, the same or similar minerals occur (sometimes) associated with weathered limestone intruded by igneous rock forming a smithsonite-limestone breccia. Such breccias are often the ore, with cavities in them being lined with botryoidal surfaces of smithsonite and hemimorphite at times delicately colored with trace elements making blue, green, or pink specimens that are quite attractive. Many of these are similar in appearance to the minerals once found at Granby, Missouri, and even the earlier mines of Valles Mines, Louisiana Territory, in what is now southeastern Missouri. Specimens of these minerals from Mexico are readily available and affordable to the geo-collecting community.

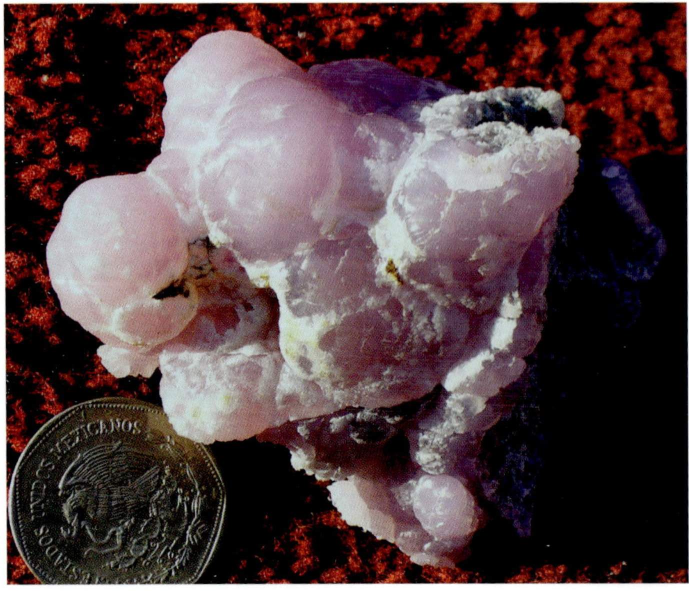

Pink smithsonite, Choix Sinaloa, Mexico. The primary sphalerite occurrence from which this smithsonite came was formed as a hydrothermal deposit, but the secondary smithsonite is similar to that found in MVT occurrences of the Ozarks. Pink smithsonite is rare, its color originating from trace amounts of cobalt or manganese.

Colorless smithsonite, Choix Sinaloa, Mexico.

Blue smithsonite, "Kelly Mine," Socorro County, New Mexico.

Green smithsonite. San Antonio el Grande Mine, Santa Eulalia, Mexico. The green color originates from trace amounts of copper. This smithsonite formed on a smithsonite conglomerate, a characteristic of most smithsonite including that from the Ozarks. (Value range F.)

Blue smithsonite, Agdalena mining district (Kelley Mine), Socorro County, New Mexico. This blue smithsonite is associated with geologic conditions (intrusions into granite), which is the geological environment also present in the Mexican smithsonite occurrences previously shown. Primary mineralization from which this smithsonite formed was hydrothermal, although the formation of secondary minerals like this was similar to that of the secondary minerals of the Ozarks.

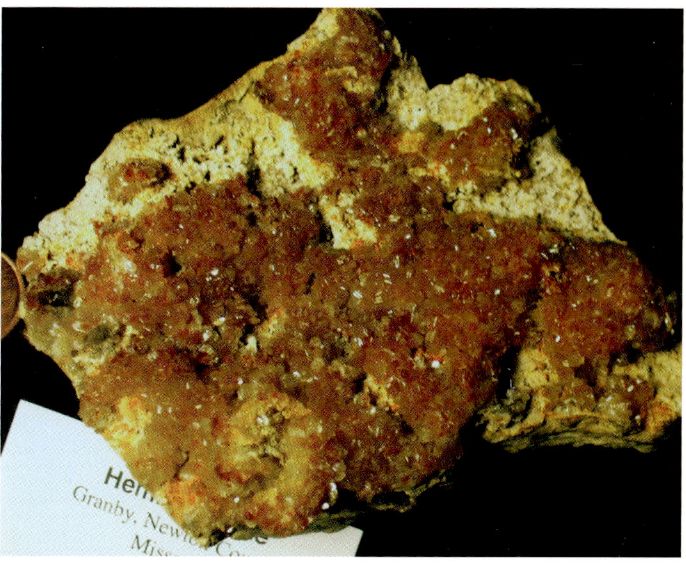

Granby hemimorphite. All of these specimens from Granby, Missouri came from old collections–they are quite collectable today. Hemimorphite is a zinc silicate. Clusters like this were associated in quantity with some of the earliest large-scale mining activity of the Tri-State district in the late 19th and early 20th centuries. (Value range E.)

Hemimorphite from Granby, Missouri. Attractive specimens of hemimorphite and other secondary or supergene minerals that formed above the water table from weathering of primary sulfides. These were mined extensively during the last third of the 19th century and early in the 20th. (Value range E.)

Blue hemimorphite. Hemimorphite often is difficult to distinguish from smithsonite–both being secondary zinc minerals. The blue color comes from trace amounts of copper, a frequently found component associated with zinc and which sometimes produces beautiful blue smithsonite as well. This specimen is not from the Granby area although similar specimens were found there when mining was active. The specimen came from China. (Value range E.)

Cluster of exceptionally large, hemimorphite crystals. Ojuela Mine, Mapimi, Dirango, Mexico. (Value range F.)

Group of clear hemimorphite crystals. Ojuela Mine, Durango Mexico.

Clear, colorless hemimorphite crystals. Hemimorphite similar to this has come from some areas of southwestern Missouri, which produced secondary lead and zinc minerals–but these are rare. Ojuela Mine, Mapimi, Durango Mexico. (Value range F.)

Elmwood, Tennessee

Some of the most striking specimens of sphalerite, fluorite, and galena found recently come from MVT mineralization in Middle Ordovician Limestone exposed in part of the Nashville Dome, an uplifted area in central Tennessee, somewhat like the Ozarks, but smaller. These gemmy specimens generally are labeled as coming from Elmwood, Tennessee, the nearest town to the mines.

Gemmy sphalerite from Elmwood, Tennessee. Beautiful gemmy crystals of sphalerite (and other minerals have come from lead-zinc mines in central Tennessee). This mineralization is associated with the Nashville Dome, an uplifted area of early Paleozoic rocks somewhat similar to the Ozarks. (Value range F.)

Fluorite perched on sphalerite, Elmwood, Tennessee

Another gemmy sphalerite crystal group from Elmwood, Tennessee. (Value range F.)

Fluorite cubes perched upon sphalerite are a frequently seen habit of Elmwood minerals. (Value range F.)

Fluorite on sphalerite, Elmwood, Tennessee. (Value range E.)

Austinville, Virginia Sphalerite Occurrence

Sphalerite and galena were mined for a number of years from Lower Cambrian limestone near Austinville, in the Appalachian Mountains of western-most Virginia. The occurrence is that of a typical MVT deposit; however, the limestone beds containing the ore are fractured and dipping at a twenty-five-degree angle. Other similar zinc mining regions occur in the southern Appalachians in eastern Tennessee. Sphalerite and galena at Austinville was mined from what is a typical MVT occurrence in the Appalachian mountains where mineralization was in a Lower Cambrian age limestone, known as the Shady Limestone. Sphalerite found by the author there was minimal and unexciting; however, the objects of interest were the peculiar archeocyathids found in the limestone excavated during mining. Austinville, Virginia was also named after Moses Austin, who, in the late 18th century, became involved with lead mining in southwestern Virginia. His interest in lead mining led Austin to the lead mines of Mine Au Breton in Upper Louisiana. His son, Stephen Austin, became one of the founders of what would become the state of Texas.

Archeocyathids from Shady Limestone zinc mines, Austinville, Virginia. These problematic fossils are found only in the Lower Cambrian. Archeocyathids resemble corals–which they are not. They appear to be related to sponges, but are usually placed separate from sponges and are considered as one of the peculiar extinct life forms that lived in the Cambrian oceans. Their lack of spicules prevents them from being undoubted sponges and they may be some sort of "experimental" organism that evolved at the beginning of the Paleozoic Era and then went extinct before the end of the Cambrian Period. They are just one example of some of the "weird" life forms characteristic of the Cambrian Period.

Foreign MVT Sphalerite Occurrences and Associated Fossils

Some excellent examples of minerals as collectables have come from MVT deposits in limestone which crops out in the Carpathian Mountains of Romania. Here, as in the Appalachian Mountains, the mineral-bearing host limestone has been tilted by tectonic forces, but otherwise the occurrences are low temperature, stratiform deposits associated with limestone.

Sphalerite with calcite and pyrite, Romania. The sphalerite is a variety that contains a very low iron content. Thick Paleozoic limestone sequences in the Carpathian Mountains of Romania host MVT mineralization like this. As in the states, sphalerite and galena predominate. Caves developed in this same limestone, besides sometimes showing mineralization, have been the source of many Pleistocene (ice age) fossils that have gone onto the fossil market, which includes the bones of cave bears.

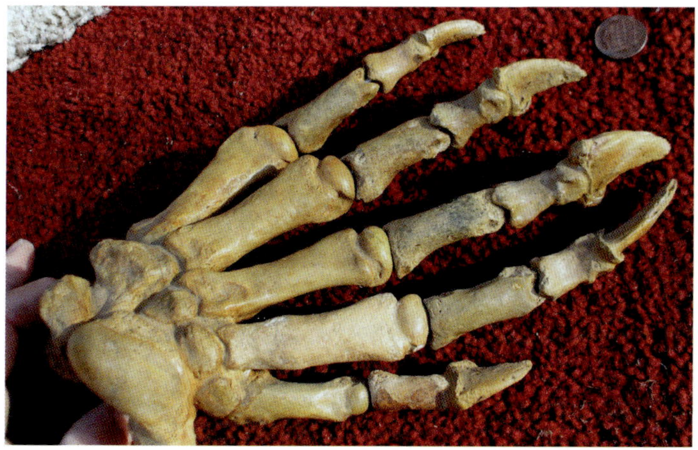

Ursus spelaeus (cave bear). Caves can trap animals which wander into them and die. They also preserve the animals' bones very well. Thick limestone beds that harbor MVT minerals, like lead and zinc, can be full of caves, as is the case in the Ozark Uplift. These fossil bones of cave bears come from caves in the Carpathian Mountains of Romania. Numerous bones of cave bears recently appeared on the fossil market and have been widely dispersed among collectors. Concentrations of cave bear bones in some of the Romanian caves were once so abundant that they were mined for fertilizer in the 19th century.

Layered smithsonite on galena. A mix of primary and secondary minerals from MVT deposits in limestone. Carpathian Mountains, Romania.

Connections between thick carbonate rock sequences—MVT minerals and caves with their fossil vertebrates—are found in the Ozarks, as well as in the Carpathian Mountains. Here is a fossil deer jaw from a cave deposit uncovered during housing construction. (Value range F.)

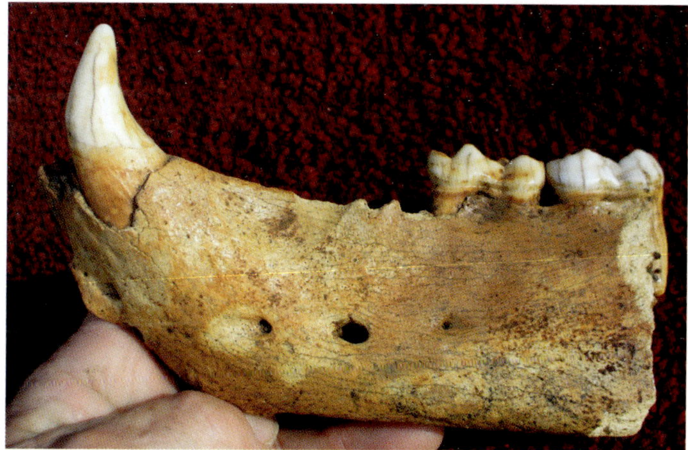

Part of the lower jaw of a cave bear from the Carpathian Mountains.

Sphalerite with a low iron content perched on quartz druze. Most sphalerite contains some iron substituting for zinc. This sphalerite from China has an unusual orange-yellow color. Leng Shuijiang Quarry, Hunan Province, China.

Close-up of low-iron Chinese sphalerite. (Value range F.)

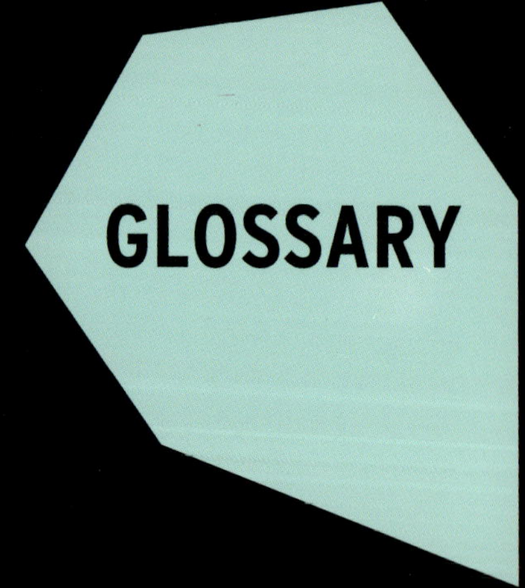

GLOSSARY

Botryoidal surface

A distinctive "bubbly"-looking surface commonly associated with smithsonite and other supergene minerals, also known as a reniform surface.

Cadmium

Cadmium is element number 48 on the Periodic Table. It is a relatively rare element. Often consistently associated with sphalerite, zinc mining is the major source of this element and the Missouri tri-state region has produced much of the worlds supply. Cadmium is the source of the yellow color in the Rush Arkansas and other smithsonite occurrences from the Ozarks.

Galvanizing

The process of coating steel or iron with a thin film of metallic zinc by immersing the iron object into molten liquid zinc. Until the use of zinc as a major component in electric batteries (dry cells), galvanizing was the major use for the element.

Germanium

An element in the same group of the periodic as carbon and silicon, germanium like cadmium is associated with sphalerite in the Kansas, Missouri, Oklahoma tri-state region. This mining region has been a major source of it. Germanium was extracted from sphalerite when it was smelted and the Eagle Picher Company stockpiled the element—an element which became quite valuable in electronics during the 1950's and '60's as the first widely used semi-conductor used germanium in transistors. The 1950's "transistor sister" radio utilized germanium rather than silicon which today is the semi-conducting element generally used in solid state electronics.

Prospects

Areas where mineralization is present, but which may not be rich enough or large enough to mine. After locating a mineralized area, the process of digging a prospect pit, or a drilling program, is used to determine the extent and richness of the mineralized area.

Supergene or secondary minerals

A mineral occurrence where primary minerals like galena and sphalerite have been subjected to weathering and oxidization, forming either sulfates, carbonates, or other oxidized forms of desirable elements. These minerals are also associated with quantities of hydrated iron oxide (ocher), which formed from oxidization of iron sulfide and known as gossan or as an "iron hat."

Chapter Five Resources

Buckley, Ernest R. and Henry A. Buehler, 1905. *The Geology of the Gran Area*. Missouri Bureau of Geology and Mines, 2nd Series, Vol. 4.

Siebenthal, C. E., 1916. *Origin of the zinc and lead deposits of the Jop region, Missouri, Kansas, Oklahoma*. United States Geological Sur Professional Paper 606.

CHAPTER SIX

COPPER, COBALT, NICKEL, AND IRON MINERALS

Copper Minerals, Joplin, Tri-State Area

Copper minerals are not as common in MVT deposits as are those of lead and zinc. They do occur, however, and it is possible that large MVT deposits may yet be discovered somewhere in the U.S. Midwest. The most widespread MVT copper mineral in collections is chalcopyrite, found as small crystals on dolomite from the Joplin, Tri-State area. Chalcopyrite is copper sulfide and is the most commonly seen primary copper mineral. Second in frequency of occurrence is malachite, often an alteration product of chalcopyrite.

Copper-bearing water, which comes from mines in Butte Montana, 1956. This mineral-laden water could be compared to that traveling through rock strata in the geologic past. When a suitable environment was encountered, copper would precipitate from the water and be deposited as a copper mineral deposit or ore body.

Utah porphyry copper mines. Much of the world's copper comes from what are known as porphyry copper deposits, which are deposits in igneous rock through which copper mineralization is widely dispersed. Copper mineralization also occurs with MVT deposits, however, unlike occurrences of lead and zinc, MVT copper occurrences are of minimal significance. This is because porphyry copper deposits, although of low copper content per unit volume of rock, can enable mining and processing on a large scale, which ultimately can be more economical, involving the so-called economy of scale.

Water charged with copper like that on the previous page being deposited electrolytically on a steel soda can. The copper will plate out on more chemically reactive minerals like iron as has happened on the surface of this steel can. Here is a mechanism or suitable environment, which precipitates a heavy metal element. There is a similarity here to MVT mineralization with its copper being deposited as a solid from copper bearing solutions. In MVT mineralization sulfur in carbonate rock and shale substitutes for the steel can.

Chalcopyrite. Copper sulfide in some instances can be an important mineral in large MVT deposits like those previously mined in the Tri-State region near Joplin, Missouri, or in Missouri's Viburnum Trend. Copper minerals can acquire a colorful tarnish as can be seen here. Brushy Creek Mine, Viburnum Trend, Reynolds County, Missouri.

Chalcopyrite crystals with a slight purple tarnish formed on dolomite, Brushy Creek Mine, Viburnum Trend.

Chalcopyrite, Brushy Creek Mine, Viburnum Trend. Copper sulfides, like chalcopyrite, often take on a tarnish, which can be bluish or purple as is the case here. This sometimes is done artificially to make the mineral more appealing and salable. (Value range F.)

Chalcopyrite crystals on dolomite, Tri-State district (Missouri, Kansas, Oklahoma). Tri-State chalcopyrite usually is in the form of small tetrahedrons perched on light pink dolomite. This is generally found associated with chert. A chert substrate is a sure sign that a specimen like this (or one of galena or sphalerite) came from this mining district. (Value range F.)

Chalcopyrite, Brushy Creek Mine, Viburnum Trend. (Value range F.)

Chalcopyrite crystals scattered over sphalerite, Tri-State (Missouri, Kansas, Oklahoma). (Value range F.)

Copper Minerals from Current River Region

Chalcopyrite and chalcocite, along with secondary malachite, was mined in the 19th century from dolostone, which is in contact with buried Precambrian knobs associated with a profound unconformity. This association of chalcopyrite occurring in carbonate rocks lying upon Precambrian knobs composed of felsic igneous rock is also found in the St. Francois Mts. Region.

This bluff of Cambrian age dolostone on Missouri's Current River is just upstream from the Slater Copper Mine, a mid-19th century mine that produced fair amounts of MVT copper as chalcopyrite and malachite. The ore body was in rubble zones, which occur just below this massive dolostone at the interface between the dolostone and underlying rhyolite.

This small outcrop of igneous rock (rhyolite) on the Current River is near the area where the Slater Copper Mine was located. Rubble zones of dolomite fragments that lie above this hard igneous rock of Precambrian age (Precambrian-Cambrian contact zone) were cemented together with chalcopyrite, the ore of this mine located along the river in the 1840s and '50s. This outcrop is just upstream of where the Jacks Fork River joints the Current River–now in the Ozark National Scenic Waterways.

Eastern Ozarks Copper Mineralization

Copper as both chalcopyrite and chalcocite has been mined in the eastern portion of the Ozark Uplift in both Jefferson and St. Genevieve counties. Here copper mineralization is (or was) associated with porous dolomite containing stromatolites. Valle Mines, named after Felix Valle, the Governeur Generale of Upper Louisiana under France in the late 18th century, yielded copper, zinc (as smithsonite) and lead (as both galena and anglesite) These ores were close to the surface and most were secondary or supergene deposits where the copper occurred as carbonates,

Malachite and calcite. Secondary and highly oxidized malachite from Valles Mines associated with iron oxide–both minerals were derived from the near-surface weathering of chalcopyrite and pyrite respectively. The iron oxide is part of what is known as gossan or "iron-hat," material, which forms from weathering and oxidization of primary sulfides.

Azurite. Azurite is a blue form of copper carbonate that is usually associated with malachite. It is rarer than malachite in most cases, occurring as a secondary mineral formed from the weathering and oxidization of primary copper sulfide minerals like chalcopyrite. It can be associated with secondary (or supergene) copper deposits formed from weathering of primary sulfide minerals.

such as malachite and azurite. A greater variety of secondary minerals were found and mined from supergene deposits in southwestern Missouri near the town of Granby; these included a variety of colorful copper minerals derived from the chalcopyrite crystals, which often coat the dolomite associated with the primary ores of the Tri-State mineral region.

Malachite in calcite geode. Geodes are peculiar rocks, often filled with crystals, in this case calcite. The dark spherical blob is malachite, a mineral rarely found in geodes. (Value range F.)

Malachite after chalcopyrite, Valle Mines, Missouri. Shallow, MVT deposits can contain copper mineralization, which can be converted by weathering to secondary minerals like malachite and smithsonite. Valle Mines was one of the early mining areas worked by the French in the 18th century in North America where mining activity took place well before the Louisiana Purchase. The major mineral mined here was galena, copper minerals taking on a secondary importance.

Conichalcite and aurichalcite, Ojuela Mine, Mapimi, Durango, Mexico. Both of these related secondary (or supergene) copper minerals were found at Granby, Missouri, as well as at other places where primary copper minerals, like chalcopyrite, underwent oxidization and weathering. Usually, the Tri-State supergene occurrences were small and few minerals were collected. Better and more accessible specimens of these minerals come from other areas—many from Mexico. This specimen is on smithsonite, a gangue mineral also associated with secondary minerals. The aurichalcite is the lighter green mineral at the right. Conichalcite is the darker green mineral. (Value range F.)

Malachite globule on barite. Small amounts of secondary copper minerals can be somewhat common with MVT mineralization—the copper usually being found in only relatively small amounts. The small malachite sphere shown here is perched on barite.

Conichalcite with limonite (iron-hat). Ojuela Mine, Mapimi, Durango, Mexico. (Value range F.)

Conichalcite. Small crystals of this secondary copper mineral from main lode, Great Australian Mine, Queensland, Australia. These secondary copper minerals occur widely where they are usually associated with other secondary minerals, like hemimorphite and smithsonite. (Value range G.)

Acicular (radiating) crystals of conichalcite. Ojuela Mine, Mapimi, Durango, Mexico.

Malachite coating a dolomite chunk. Copper as chalcopyrite, cuprite and malachite was mined in the eastern Ozarks of Ste. Genevieve County early in the 20th century (Cornwell Copper Mines). Most of what is found on the mine dumps today is malachite, a mineral which often cements together chunks of dolomite into a conglomerate. This deposit appears similar to deposits to the north at Valle Mines. (Value range F.)

Linarite. A typical specimen of the thin crystals of this intensely blue mineral that commonly coats quartz. Blanchard Mine, Bingham, New Mexico.

Linarite. This mineral was reported from Granby (Beer Cellar Mine) along with other occurrences of secondary minerals in the early days of mining in the Tri-State district. Specimens generally were small. This is a specimen from New Mexico where spectacular examples of this intensely blue copper mineral are found in the Blanchard Mine, Bingham, Harrisonburg District, Socorro, New Mexico. (Value range F.)

Fossil gastropods (snails) from Cornwell Copper Mines, Ste. Genevieve Co., Missouri. This dolomite slab has some of the green stain of malachite. (Value range F.)

Cobalt Minerals

The southern portion of the old lead belt, in the Mine LaMotte and Fredericktown areas, carries concentrations of cobalt minerals—primarily in the form of the relatively rare cobalt mineral siegenite. A cobalt-nickel sulfide, siegenite occurs as pinkish or brownish crystals with a metallic luster—they somewhat resemble galena and in MVT deposits are always associated with that mineral. The host rock of siegenite is dolomite of the Bonneterre Formation of Middle and Late Cambrian age. Usually siegenite crystals are small, looking somewhat between pyrite and galena. Mine LaMotte, a MVT deposit rivaling Mine au Breton (in its early discovery) has siegenite as one of its sulfide minerals associated with both chalcopyrite and galena.

Micromount of siegenite crystals, Buick Mine, Iron County, Missouri. (Value range F.)

Siegenite. Siegenite is cobalt sulfide. It's a relatively rare mineral found only at a limited number of locations. This is a slab covered with siegenite crystals which come from the southern portion of the Virburnum Trend. (Value range F.)

Siegenite crystal group arranged in a ring. Sweetwater Mine, southern portion of Virburnum Trend, Missouri. (Value range F.)

Group of siegenite crystals and other sulfides with associated Viburnum Trend crystals. *Glenn Williams collection.*

Octahedral galena and calcite crystals fill a vug in this Mine La Motte specimen. Siegenite occurs above the X.

Galena (grey), pyrite, and siegenite. Small crystals of siegenite are associated with the pyrite; this type of mineralization was extensive at the Mine Lamotte-Fredericktown deposits making them a source of cobalt. X marks the areas of siegenite. (Value range G.)

Siegenite, with its cobalt, is in the galena surrounded mass left of center. Mine Lamotte, Fredericktown, Missouri.

Above the X is an intergrowth of pyrite and siegenite. Fredericktown (Mine Lamotte), Missouri, occurrence.

Cobaltocalcite (or cobaltian calcite) calcium carbonate containing small amounts of cobalt. This is found as a secondary mineral associated with cobalt ores like that which occur in portions of the old lead belt. Specimens similar to this have recently come from MVT occurrences in Morocco. Bou Tzzer, Tazenakhi, Morocco. (Value range F.)

Another Moroccan specimen of cobaltian calcite.

Cobaltian calcite. Democratic Republic of Congo. (Value range G.)

Nickel Minerals

The nickel minerals siegenite and bravolite are found in the Fredericktown-Mine LaMotte area associated with siegenite and chalcopyrite. The other MVT occurrences of nickel minerals are millerite and its oxidization products. Millerite is nickel sulfide and its occurrences usually are entirely separate from other MVT deposits. This isolation of millerite from other sulfide minerals has been suggested

Millerite needles, which occur along fractures in flinty chert. Millerite is relatively common in some flinty-chert layers near Troy, Missouri. It occurs as flattened needles, which formed along very small and inconspicuous fractures in chert.

Millerite "hair" in geode. Clusters of fine, hair-like crystals of millerite can be found within geodes—mostly in northern Missouri, southeastern Iowa, western Illinois, and Kentucky. The millerite in these geodes resembles steel wool. (Value range F.)

as a consequence of nickel being derived from an extraterrestrial source. Nickel is an element commonly associated with meteorites and nickel in a once-existing meteorite field, which then weathered and was mobilized into underlying limestone layers is the extraterrestrial source. In other words, weathering of a strewn field of meteorites sometime in the geologic past may have been the source for the nickel in millerite. Nickel, which once put into a soluble form from weathering, migrated with descending groundwater into underlying limestone, where it reacted with sulfur present in the limestone to produce millerite.

Millerite needles in calcite, Troy, Missouri.

Millerite "needles" embedded in calcite. Calcite cleavage masses shown here contain millerite needles that are now embedded in this translucent mineral. Troy, Lincoln County, Missouri. (Value range E.)

Close-up of spray of millerite needles. Troy, Missouri.

Pecoriate in quartz geode, St. Louis County, Missouri.

Pecoriate in quartz geode. This nickel mineral formed as an alteration product of millerite. Like many secondary nickel minerals, pecoriate is green, but it's a different, paler green from secondary copper minerals like malachite. Fern Glen Formation, I-44 excavations, Antire Road, St. Louis, Missouri area. (Value range E.)

Hoenessite. Another secondary nickel mineral derived from millerite needles. Troy, Missouri. (Value range D.)

Iron Minerals

Iron minerals are ubiquitous in most natural deposits, which of course include mineral deposits. Iron would be expected to be abundant, as it is one of the most abundant elements in the earth's crust. The sulfide minerals pyrite and marcasite can be collectable iron minerals, although the latter is often prone to decompose in the presence of water and oxygen and then to disintegrate. Pyrite (fool's gold) occurs commonly in many rocks, marcasite being less common—being a mineral that forms in a very reducing environment and is often stable only in this kind of chemical environment. Some of the most collectable marcasite specimens are the cockscombs found associated with galena in the Tri-State region. This marcasite is not only stable, but attractive—a characteristic of many other occurrences of this mineral being that it can be ugly.

Marcasite has a crystalline structure different from that of pyrite, although chemically it is the same as pyrite. Marcasite is also less stable than pyrite, specimens often decomposing over time in a collection or in the field. Marcasite from the Tri-State area near Joplin, Missouri like this are usually stable. (Value range F.)

Marcasite, Viburnum trend. Marcasite, like pyrite, is a frequently occurring mineral. Like marcasite from the Tri-State region, that of Missouri's new lead belt appears to be relatively stable. (Value range F.)

Pyrite, Viburnum Trend. Pyrite is a common mineral in most MVT occurrences. It's one of the most abundant sulfide minerals. (Value range F.)

Marcasite "suns." These disks of iron sulfide form and come from shale above a mined coal seam in southeastern Illinois. Large numbers of them have been collected by coal miners from the roofs of coal mines near Sparta and Freeberg, Illinois. (Value range F.)

Pyrite and calcite filled vug. St. Louis, Missouri area. (Value range F.)

Marcasite crystals on sphalerite. Tri-State region, Joplin, Missouri. (Value range F.)

Pyrite crystals in a limestone vug. Cavities in rocks often can be lined with crystals. Here, small pyrite crystals cover the vugs quartz-lined surface. (Value range G.)

Pyrite mass made of crystals. Masses like this often occur in shale where they may have formed not too long after the shale was deposited as clay. St. Louis, Missouri area. (Value range G.)

Iron Mine, (goethite and pyrite), south of Rolla, Missouri.

Goethite "pipes" set into a wall. Iron oxide can form in many peculiar shapes, a common one being elongate "pipes," which resemble stalactites and known as stalactitic or pipe iron ore. These usually occur in near surface deposits, usually in residual clay formed (or left behind) from deep weathering of limestone or dolomite.

Botryoidal goethite (or limonite) Eleven Point River, S. Missouri.

Pyrite "acorn." Pyrite can form concretions and concretionary masses of many different shapes. These are often associated with shale layers, which have no other minerals (but may contain fossils). Notice the cubic crystals at the top of the acorn--a characteristic crystal form of pyrite.

Limonite or goethite pseudomorphs after marcasite. These iron oxide minerals are common in the Ozarks, being widely distributed over the uplift. Originally, the crystals were marcasite (iron sulfide) that has been oxidized to goethite or limonite. The crystal structure of marcasite has been retained–that is what a pseudomorph is: a "false form" of a crystal.

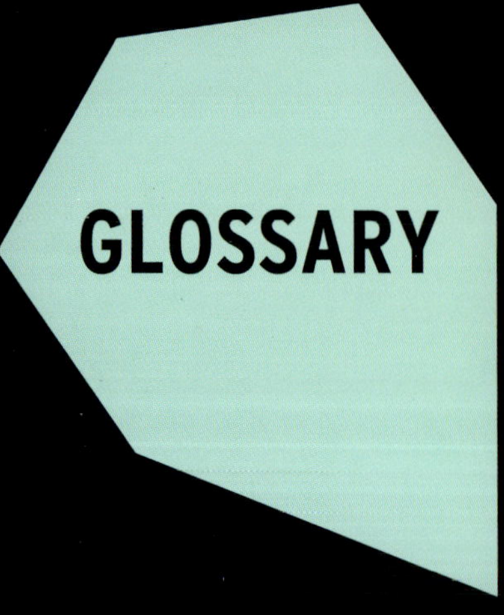

Extraterrestrial origin

The origin of something not being from the earth—in reference to some nickel minerals found in a MVT setting. Specifically, this refers to nickel minerals, like millerite and its alteration products, such as pecoraite, which are found either in limestone or in geodes that come from beds of limestone in the U.S. Midwest.

Iron, colbalt, and nickel

These three elements occur together on the periodic table. They appear to be some of the highest atomic number elements produced in a nova which is believed to be the source of them. All three are siderophile elements which means that in a liquid or gaseous state they have an attraction for iron. The majority of these three useful elements in the earth reside in the Earth's core.

Reducing chemical environment

A chemical environment opposite to that known as oxidization. Sulfide minerals, like chalcopyrite and sphalerite in an MVT setting, are associated with reducing conditions and form in a reducing chemical environment. Secondary minerals, like malachite and smithsonite, derived from these minerals are formed in an oxidizing chemical environment associated with weathering of the original chalcopyrite and smithsonite.

Siderophile Elements

These are elements which are naturally associated with iron and include the vast amount of cobalt and nickel lodged in the earth's partially molten core. Other less common elements like platinum, iridium and osmium are also siderophile elements and most of these are also sequestered in the Earth's core.
Unconformity

A major break or hiatus in a sequence of rocks. Specifically refers to beds of fragmented rock found to bear copper minerals in the central Ozarks of Missouri.

Chapter Six Resources

Bain, Harry F. and E. O. Ulrich, 1905. *The Copper deposits of Missouri.* U.S. Geological Survey Bulletin 267, 52 pp.

Bridge, J., 1930. *The Geology of the Eminence and Cardareva Quadrangles.* Missouri Bureau of Geology and Mines, 2nd Series.

Grawe, Oliver R., 1945. *Pyrite deposits of Missouri.* Missouri Geological Survey and Water Resources. Vol. 30, 2nd Series.

Haworth, Erasmus, 1886. *Millerite*, Science Vol. 8, p. 369.

Keller, Walter D., 1940. *Aurichalcite (basic zinc copper carbonate) in Missouri.* American Mineralogist, Vol. 25, No. 5. pp. 375-376.

Keys, Charles Rollin, 1903. *Significance of the occurrence of minute quantities of metalliferous minerals in rocks.* Iowa Academy of Science Proceedings. Vol. 10, pp. 99-103.

Le Font, 1984. *Siegenite from the Buick Mine, Bixby, Missouri.* The Mineralogical Record, Vol. 15, No. 1, pp 37-39.

Pirssion, Louis V. and H. L. Well, 1894. *On the occurrence of leadhillite in Missouri and its chemical composition.* American Journal of Science

CALCITE AND DOLOMITE

Calcite

Calcite, the major component of limestone, might be expected to be a common mineral found in MVT deposits. This is a no-brainer, as MVT minerals are usually associated with limestone or with the related rock dolostone. Dolomite crystals of a distinctive habit and color (light pink), like that associated with the Missouri Tri-State district, are well known. Large numbers of these dolomite crystals have been dispersed worldwide with their toppings of small sphenoids of chalcopyrite crystals. More intensely pink dolomite crystals are associated with northern Arkansas occurrences, those from quarries near Black Rock, Arkansas being especially noteworthy.

A display of calcite crystals in a serious collectors museum. The crystals are from Missouri's Viburnum Trend. Calcite occurs in a variety of habits, complete coverage of which could fill an entire book. *Glenn Williams collection, 2009.*

Calcite crystal from cavity occurring in the previously shown outcrop.

A recently exposed mass of calcite exposed in a new road cut on U.S. Highway 67, near Fredericktown Missouri. Large masses of calcite are fairly common in limestone and dolostone outcrops, especially fresh ones like this. Cavities associated with such outcrops are often lined with crystals, although these may or may not be collectable.

Fractured dolostone filled with calcite, Piedmont, Missouri. Fresh road cut on Missouri Highway 34.

Calcite crystals from near Piedmont, Missouri.

Tri-State Mining District Calcite

Calcite varies considerably in its crystal habit and in its various crystal forms can be a geo-collectable. Many of these have come out of the mines of the Tri-State area near Joplin, Missouri; Picher, Oklahoma; and Galena, Kansas. Missouri's new lead belt also produces a variety of spectacular calcites.

Group of hobnail calcite crystals, Joplin, Tri-State region. (Value range E.)

Hobnail calcite crystal. This is one of the more distinctive habits of calcite. Viburnum Trend, Missouri.

Large dogtooth calcite crystal from Joplin, Tri-State region. This came from one of the "crystal caves," large calcite lined cavities that occur associated with the Tri-State mining region, but now all are underwater, as they lie beneath the water table.

Average Joplin Tri-State crystal group. The dark crystals are sphalerite. The crystals are attached to chert, a characteristic of most Tri-State minerals. (Value range F.)

Calcite from Vugs/Geodes of the Warsaw Formation

The Mississippian age (~330 million years old) Warsaw Formation is noted in the U.S. Midwest for both its geodes and its crystal-filled vugs. Geodes are often round, hollow, crystal-filled rocks that have a characteristic surface that enables a trained eye to easily spot them—those of the Midwest often weathering from limestone and siltstone of the Warsaw Formation. Numerous cavities (vugs) also occur in the Warsaw Formation and these can be lined with gemmy calcite crystals. Here are some examples of these collected from excavations in the northeastern Ozark region southwest of St. Louis.

Honey calcite covered with clear calcite in a geode/vug. (Value range F.)

Clear and white dogtooth crystals in vug.

Rhombohedral crystal from geode/vug.

Honey calcite vug/geode.

Brown, iron-bearing rhombohedral calcite with clear calcite. The brown crystals resemble fluorite and were thought to be such when collected. (Value range F.)

Group of rhombohedral crystals from same zone as the above.

White dogtooth calcite in vug/geode.

White dogtooth calcite with sphalerite crystal at right.

Clear or white dogtooth calcite with pyrite specks.

Honey calcite vug/geode.

Calcite from Missouri's Viburnum Trend

A variety of calcite habits are associated with the Viburnum trend—Missouri's new lead belt. One of the most distinctive are elongate prisms with terminations in a rhombohedron. These are especially characteristic of the Sweetwater Mine in the southern portion of the trend.

"Poker chip" calcite (hobnail) from a quarry in the southern part of the Viburnum trend in Shannon County, Missouri.

These elongate prismatic calcites are typical of the Sweetwater Mine, Viburnum Trend. (Value range F.)

Calcite globs on marcasite, Viburnum trend, Missouri.

Dogtooth spar group on octahedral galena. Brushy Creek Mine, Viburnum Trend. (Value range E.)

Calcite, Brushy Creek Mine, Viburnum trend.

These elongate prismatic crystals are a common form of calcite from Missouri's Viburnum Trend, especially from the Sweetwater Mine.

Calcite and Siderite from Large Black Concretions

In the U.S. Midwest, thin beds of limestone and shale often overlie a seam of coal that has been mined both in strip mines and underground mines. One place where these concretions were found by the author is along the Missouri River at a spot noted by explorer William Clark in the journal of their famous 1804 expedition. These black concretions were associated with weathered shale, which overlies a mid-Pennsylvanian coal seam exposed during the time of their expedition, but now covered with talus. These black concretions are better known associated with the same coal seam mined by strip or surface mining. The concretions are heavy, being composed of siderite (iron carbonate). The interior of the concretions have what resemble shrinkage cracks lined with both brown crystals of siderite and sprinkled with clear calcite—an attractive and striking mineral combination. These mineral-bearing concretions were more frequently available a few decades past as surface mine reclamation now requires almost immediately burial of overburden and covering it with soil.

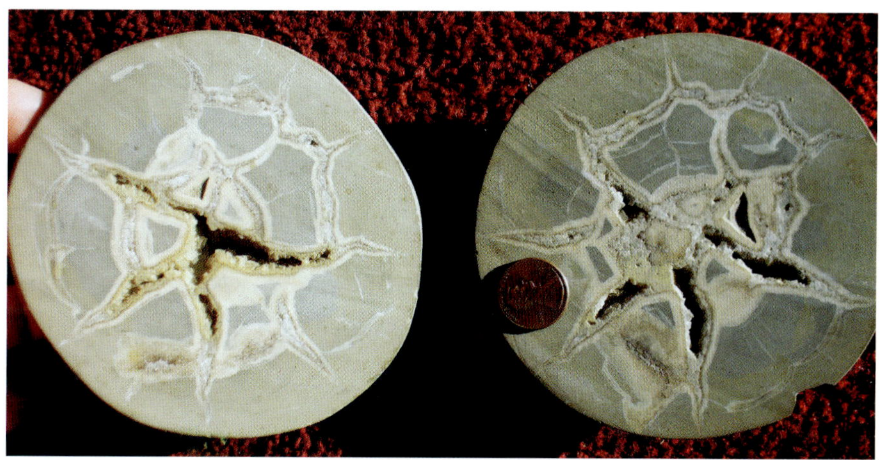

Sliced concretion similar (but smaller) to those that produce the above siderite and calcite crystals. These crystals grow in what appear to be shrinkage cracks. (Value range F.)

Group of small crystals which grew in a "crack" in a large, black concretion. Freeberg, Illinois. (Value range F.)

Calcite crystals on black limestone. This came from a large black concretion found along the south side of the Missouri River about twenty one miles above its confluence with the Mississippi. This is the same place noted by William Clark (of the famous 1804 Lewis and Clark Expedition) at what he called "Coal Hill" and the French called "Carbonere" (Charbonier Road in N. St. Louis County is also named for this). Clark notes the following in his journal "this hill appear(s) (sic) to contain (a) great variety of coal (and ore of a appearance) etc." Clark's "ore" may have been in reference to these black concretions that occur just above the coal bed, which no longer is exposed at this place. The black concretions are evident during low water where they have weathered from shale, which occurs above the coal bed. The period half-cent coin is the size of a modern quarter.

Siderite crystals (upon which is perched a calcite crystal) from a large, black concretion occurring above the same coal seam noted by Clark along the Missouri River. Freeberg, Illinois. (Value range F.)

Calcite on siderite, southeastern Iowa. These crystals have also come from large, black concretions associated with strata overlying a coal seam. (Value range E.)

St. Louis Area Crystals

Extensive quarrying in limestone beds of the St. Louis area at times encounter concentrations of calcite, dolomite, fluorite, and millerite crystals. Here are a few of these from collections of the St. Louis Science Center.

Mixed white and honey-yellow fluorite in vug, St. Louis, Missouri. Limestone beds in most areas can contain calcite crystals. The presence of yellow fluorite in the center of this group makes it more unusual. *Courtesy of St. Louis Science Center.*

Group of calcite crystals, St. Louis Missouri. *Courtesy of St. Louis Science Center.*

Vug containing dolomite (bottom) and calcite crystals. Mississippian, St. Louis Limestone, St. Louis Missouri. *Courtesy of St. Louis Science Center.*

Calcites from Elsewhere in the U.S. Midwest other than the Ozarks

Calcite can exhibit a wide range of habits, often of a type that may be unique to a particular locality. Here are a few from the U.S. Midwest and elsewhere.

White rhombohedral calcite on honey-yellow calcite. St. Louis, Missouri. (Value range G.)

Gemmy crystals of MVT minerals such as sphalerite, fluorite, and calcite come from a MVT deposit in central Tennessee. Many of the minerals from this deposit near the town of Elmwood are very gemmy and thus are quite attractive, as well as being desirable.

Group of Elmwood dogtooth calcite crystals. (Value range F.)

Elongate dogtooth calcite on fluorite, southern Illinois fluorite district.

Pyrite and/or marcasite crystals are included in these calcite crystals that come from quarries in Silurian dolostone near North Aurora, Kane County, Illinois. (Value range G.)

Calcite containing small pyrite or marcasite crystals. North Aurora, Kane County, Illinois.

Elmwood, Tennessee calcite. (Value range F.)

Phantom within a dogtooth calcite crystal. Calcite exhibits a large variety of habits. These are often distinctive as to the locality from which they came. Anderson, Indiana.

Dogtooth calcite. Older collections often contain dogtooth crystals of this habit. Pugh Quarry, Ohio. (Value range F.)

Yellow calcite from a septarian concretion. Grey concretions found in shale beds (mentioned in Chapter Four), almost always contain this type of yellow calcite that forms on surfaces of the concretions internal "cracks."

Honey-yellow crystal from vugs in Devonian limestone near North Vernon, Indiana. Cavities and vugs in limestone often yield various types of calcite crystals. These are from a quarry in an area that otherwise is flat land with no outcrops—a rich agricultural region.

Manganocalcite. Pink calcite like this, containing small amounts of manganese or cobalt, is known respectively manganocalcite and cobaltocalcite, or cobaltian calcite. Both forms have been found in the old lead belt area where it was determined to have formed in the mines during the past 100 years from mineral laden water.

Dogtooth calcite, Pugh Quarry, Ohio.

White calcite on hemimorphite. Similar to specimens found in the early 20th century at Granby, Missouri. From Safu Mare, Maramures County, Romania. (Value range F.)

Dolomite

Dolomite, unlike calcite, does not occur in as many varieties as does calcite. Here are some examples of some collectable dolomite crystals.

Black Rock, Arkansas dolomite on which are perched calcite crystals with chalcopyrite inclusions. (Value range G.)

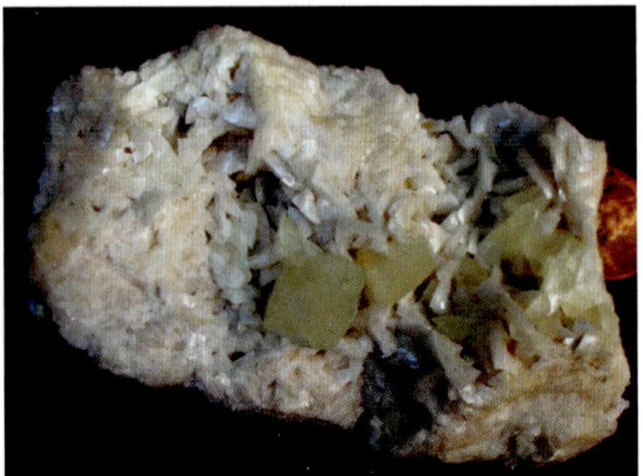

Black Rock, Arkansas dolomite crystals. (Value range G.)

Dolomite crystals, Tri-State district, Missouri. The small, dark crystals are chalcopyrite, a characteristic feature of Tri-State dolomite.

Calcite, dolomite and sphalerite, Tri-State, Joplin area. Calcite is near the top, dolomite is slightly pinkish with calcite crystals covering the dolomite. Sphalerite is the dark mineral. Small crystals of chalcopyrite usually cover dolomite crystals from the Tri-State occurrences, they are the dark spots. (Value range F.)

Dolomite crystals with a scattering of small tetrahedral of chalcopyite crystals characteristic of the Missouri Tri-State region.

Large specimen of this widely distributed form of dolomite crystals from Black Rock, Arkansas.(Value range F.)

Tri-State region dolomite with numerous chalcopyrite crystals.

Dolomite with small chalcopyrite crystals. Black Rock, Arkansas. Numerous nice crystals of pink dolomite have come from quarries in Lower Ordovician dolostone, which crops out in the northern part of Arkansas.

Calcite on dolomite, Black Rock, Arkansas. (Value range G.)

Ferroan dolomite, Park Hills Missouri. Ferrous iron can substitute for magnesium in the space lattice of dolomite–surface oxidization of the iron can then produce an iridescent tinge on the crystal. (Value range G.)

Ferric oxide-stained (pink) dolomite crystals. Potosi, Missouri. Iron in ferroan dolomite is in the ferrous oxidization state; here it is in the ferric state, which produces the red color. (Value range G.)

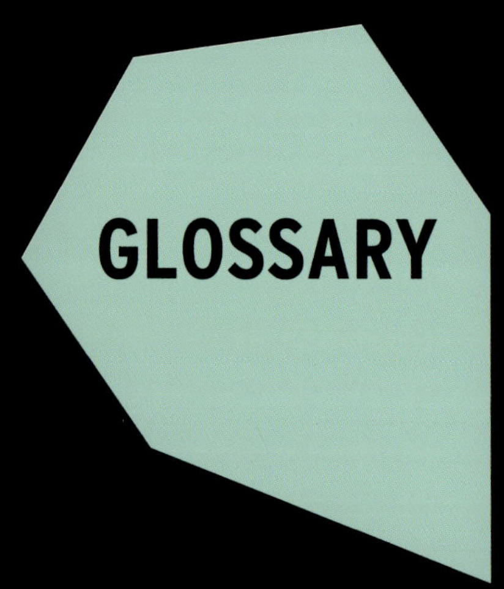

GLOSSARY

Gangue minerals

A mining term for minerals and mined material that has no commercial value and which is usually placed in "mine dumps." Calcite and dolomite are generally

Calcium.

The metallic element calcium is number 20 on the periodic table. As an element it is a brittle, reactive metal which in its elemental state has few uses. Calcium is similar to magnesium which is the lighter of the two elements (atomic number 12). Calcium and magnesium are the critical elements in limestone and dolomite.

Calcite and Dolomite.

Calcite is calcium carbonate, it's the critical component of limestone. Dolomite can be both a mineral and a rock which chemically is calcium-magnesium carbonate. The rock dolomite is sometimes known as dolostone to distinguish it from the mineral dolomite. Both limestone

FLUORITE AND RELATED MINERALS

Southern Illinois Fluorite
Rosiclair District

Fluorite, also known as fluorspar, is a halide mineral composed of the halogen element fluorine combined with calcium. One of its largest occurrences in a MVT setting is in southern Illinois and adjacent Kentucky. Here this attractive mineral occurs as replacement of Mississippian limestone (St. Louis and Ste. Genevieve Formations) in stratiform deposits associated with smaller amounts of galena, sphalerite and other MVT minerals, like barite and strontianite. A difference between this and other MVT mineral deposits does exist, for in the southern Illinois fluorite district, igneous intrusives occur along the Ohio River near Rosiclair, where major occurrences of fluorite also occur. This lends credence to the hypothesis that mineralization in this district is of hydrothermal origin and not of the same genesis as other Midwestern MVT mineralized areas, like the Missouri lead belts and the Tri-State sphalerite region. In consideration of a possible hydrothermal origin of the Rosiclair fluorite district, it is of interest to note that fluorite crystals similar to those of the Illinois district occur sporadically in the St. Louis limestone of the St. Louis area. The presence of certain minerals being associated with specific stratigraphic horizons (in this case, the St. Louis and Ste. Genevieve limestones) exemplifies a property of MVT deposits—if these occurrences were of hydrothermal origin, they presumably would not have this continuity—stratigraphic continuity of similar mineralization being one of the attributes of MVT deposits.

Cluster of relatively large purple cubes of fluorite. (Value range E.)

Cluster of fluorite cubes on calcite. Minerva Mine, Rosiclair district, S. Illinois. (Value range D.).

The color range of southern Illinois fluorite in cleavage fragments. Purple is the most common color, clear is next, then yellow. Green is relatively rare. The source of the color in fluorite is unknown. It has been suggested that it's from small amounts of hydrocarbons. This appears unlikely as colored forms of the mineral can occur in environments that involved fairly high temperatures, such as with igneous rocks.

Yellow and purple fluorite crystals and cleavage octahedron (bottom left) under transmitted light. *Glenn Williams Collection.*

Somewhat battered fluorite cubes. These are typical specimens retrieved from ore piles. (Value range F.)

Dark purple cubes. Minerva Mine, Rosiclair Ill. (Value range F.)

Same specimen as above, but shown with transmitted light.

Fluorite (dark) and calcite in vug, St. Louis Co., Missouri The Mississippian age St. Louis Limestone is the source for much of the fluorite of southern Illinois. Fluorite is found to be a fairly common mineral occurring in small quantities in the St. Louis and overlying Ste. Genevieve limestones. Why it occurs in this stratigraphic continuity is unknown. It's believed that fluorine bearing fluids traversed these limestones when they were deeply buried in the geologic past. This took place well before any of the present topography existed with the fluids moving through the limestone beds and following them for millions of years.

Pea Ridge Iron Mine and Hydrothermal Fluorite

Large, presumed hydrothermal masses of iron oxide-rich igneous rock of mid-Precambrian age occur over the Ozark region. Some of these have been mined on a large scale for iron as at Iron Mountain, Pilot Knob, or Pea Ridge south of Sullivan, Missouri. Fluorite is rare in Ozark mineralized areas—almost none occurs in the Tri-State district or the Viburnum Trend. About the only exception to this lack of fluorite in the Ozarks is its occurrence in the deep Pea Ridge Mine and the Silver Mines area near Fredericktown, Missouri.

Purple fluorite from the Pea Ridge iron mine, south of Sullivan, Missouri. Here, purple fluorite occurs associated with Precambrian igneous rock. It has all of the characteristics of the fluorite of southern Illinois. It's also one of the few occurrences of fluorite in the Ozark Uplift of Missouri.

Yellow fluorite cube. Pea Ridge iron mine, Sullivan, Missouri. (Value range F.)

Green fluorite from an undoubted hydrothermal source, Silver Mines, Madison County, Missouri. This fluorite came from a quartz vein associated with undoubted hydrothermal minerals, which includes argentiferous galena, topaz, hubernite, and mica. All of these minerals are clearly representative of hydrothermal conditions. Fluorite is the least reliable mineral indicative of MVT deposits. While found in many MVT settings, it still could have a hydrothermal origin. This fluorite and the previously shown specimens are all associated with ancient Precambrian rocks. (Value range F.)

Fluorite Fossils

MVT minerals sometimes can replace the limestone into which they have been introduced. Sometimes this will include fossils—especially in the U.S. Midwest where fluorite is associated with fossil-bearing Mississippian age limestone.

Bellerophon sissile. This is a superb specimen of a peculiar Paleozoic fossil from the Ste. Genevieve Limestone. The shell of this planispiral gastropod has been replaced with jasper. The Ste. Genevieve limestone is the major host rock for the fluorite deposits of southern Illinois.

Sliced bellerophontid with fluorite (purple) and calcite (white) replacing part of the molluscan shell. Note the inner whorl of the shell has been replaced with purple fluorite as well as jasper. Fluorite is a prevalent mineral in the Ste. Genevieve Limestone from which this fossil originated.

Another sliced specimen of Bellerophon sissile showing fluorite replacement of the shell. Ste. Genevieve Limestone, Ste. Genevieve, Missouri.

Blastoid (Pentremites) replaced with fluorite. Rosiclair flourite district, S. Illinois.

Blastoids replaced with fluorite, Southern Illinois. The rounded objects on this slab are the blastoids. Blastoids are extinct echinoderms that lived in late Paleozoic seas—especially during the Mississippian Period. *Courtesy of John Stade.*

Group of well-preserved blastoids of Missisippian age.

Elmwood, Tennessee Fluorite

Some of the most stunning fluorite specimens come from MVT deposits in central Tennessee near the town of Elmwood. Associated with Middle Ordovician limestone in the Nashville Dome, Elmwood minerals are especially unique in their attractive and gemmy appearance. Elmwood mineralization also appears to be related to the Illinois-Kentucky fluorite district of the Ohio River with its prevalence of fluorite. The age of the limestone hosting the mineralization is different, however: the southern Illinois district being in Mississippian age strata and that of Elmwood being in older strata of the Ordovician Period.

Fluorite on sphalerite, Elmwood, Tennessee. (Value range E)

Fluorite cube, Elmwood Mine, Elmwood, Tennessee.

Fluorite on sphalerite, Elmwood, Tennessee.

Elmwood fluorite cubes.

Fluorite with clump of barite. These barite clumps are characteristic of Elmwood specimens. (Value range E.)

The delicate darker border of purple is characteristic of Elmwood fluorite. (Value range F.)

Typical Elmwood fluorite specimen.

Large Elmwood fluorite cube. (Value range E)

Fluorite from the Findlay Arch of Northern Ohio and Southeastern Michigan

Brown fluorite associated with celestite and other minerals occur in northern Ohio and southeastern Michigan and have been widely distributed among dealers and collectors. These minerals come from numerous vugs and other openings in Silurian dolostone quarried there. The fluorite frequently is associated with celestite and native sulfur.

Brown fluorite crystals with a covering of elongate celestite crystals. Stoneco Quarry, Maybee, Michigan. (Value range F.)

Brown fluorite and celestite. Stoneco Quarry, Maybee, Michigan.

Dogtooth crystals perched upon brown fluorite. This habit is characteristic of minerals associated with the Finley Arch of northern Ohio and SE Michigan. (Value range F.)

These brown fluorite crystals are on calcite. Usually some dogtooth calcite is on the fluorite like the card says; however, this is not the case with this specimen. Pugh Quarry, Custer, Ohio. (Value range E.)

Purple fluorite on brown. Auglaze Quarry, Junction, Ohio.

Two generations of fluorite. Purple fluorite (2nd generation) on brown fluorite (1st generation). Findlay Arch, Auglaze Quarry, Junction, Ohio.

Some Other Fluorite Occurrences

Fluorite is a widely occurring mineral that is both attractive and variable—hence collectable. Its wide range of colors makes it especially desirable.

Fluorite from the Rogerly Mine, England, under natural light. (Value range E.)

Clear fluorite, probably of hydrothermal origin. Socorro, New Mexico.

Green fluorite, Socorro, New Mexico.

Same specimen under fluorescent light (twirley light bulb).

Same specimen under incandescent light (light bulb).

CHAPTER NINE

MVT MINERALS FROM PARTS OF NORTH AMERICA OTHER THAN THE OZARKS

Crystals from Elmwood and Gordonville, Tennessee

A MVT deposit in central Tennessee has produced attractive specimens of fluorite, sphalerite, and galena—these, in fact, are some of the most attractive MVT minerals available anywhere. Elmwood minerals stand out notably from other similar occurrences in their gemmy appearance, an attribute that makes them especially collectable and desirable.

The Elmwood occurrence is in Middle Ordovician limestone of the Nashville Dome, an uplifted structure similar (but smaller) to the Ozark Uplift of Missouri and Arkansas. Also different from Ozark sphalerite occurrences is the abundant association of fluorite. Elmwood minerals with their "gemmy" and bright appearance are really "flowers" of the mineral kingdom.

Elmwood fluorite. The purple specks and purple rim of this crystal is characteristic of Elmwood fluorite. (Value range F for each.)

Elmwood fluorite.

MVT Minerals of the Findlay Arch of Ohio and Michigan

The northern parts of Ohio and southeastern Michigan are very flat. There is little relief here with the lands glaciated surface laid out into a checkerboard of prosperous farms. Scattered over this fertile farmland are occasionally found large quarries gouged out of the earth to supply crushed rock for the numerous roads that transverse the region. Much of this rock, as crushed stone, also finds its way as aggregate in cement for urban areas like Toledo, Cleveland, and Columbus, Ohio. These quarries are in limestone and dolomite (mostly of Silurian and Devonian age), which can be quite thick, lending, because of this, to large quarrying operations. Quarries along what is known as the Findlay Arch in northern Ohio and southeastern Michigan are especially noted in their production of nice specimens of celestite and fluorite. Fractured and vug-rich portions of these carbonate rocks can sometimes contain excellent crystals of minerals typical of MVT deposits like galena, sphalerite, fluorite, and calcite. Also found are minerals like celestite and native sulfur, minerals not found in the carbonate rocks to the south in Kentucky and Tennessee or in the Ozarks. These occurrences, like those to the south, also have no connection with any igneous or related hydrothermal activity.

Brown fluorite with calcite. Pugh quarry, Custar, Wood County, Ohio.

Brown fluorite with elongate celestite crystals, Stoneco Quarry, Maybee, Michigan. (Value range F.)

Celestite and Native Sulfur

Many of the minerals found in the quarries of the Findlay Arch are different from those to the south with minerals like celestite and native sulfur occurring, as well as different forms or habits of fluorite and calcite. Another link to MVT occurrences and sedimentary rocks found in the quarries is the occurrence of asphalt and heavy petroleum. This material may coat the mineral specimens with oil, which has to be removed with a solvent. When removed, there may show beautiful, gemmy crystals unaffected by weathering as they were protected by the oil.

The occurrence of native sulfur in this area is especially interesting. Native sulfur can be associated with active volcanic regions—fumerols can deposit sulfur at the surface from gases that emanate from them. As is the case with other MVT mineralization, however, there is absolutely no indication of volcanic activity in the region of the Findlay Arch. The sulfur is believed to have come from gypsum (calcium sulfate), where the sulfur was reduced to become native sulfur by oxidization of associated petroleum (hydrocarbons). The process may have been encouraged by microbes (bacteria), which feed upon and metabolize hydrocarbons present in the rocks.

Celestite. White Rock quarry, Clay Center, Ohio. (Value range F.)

Celestite. Stoneco quarry, Maybee, Michigan. Celestite is rare to almost non-existent in Ozark MVT occurrences. (Value range F.)

Celestite, Stoneco quarry, Maybee, southeastern Michigan. (Value range F.)

Calcite crystals containing numerous pyrite inclusions from northern Illinois west of Chicago, (North Aurora, Illinois). Clusters of these crystals from a large quarry have recently (2012) shown up in quantity in rock and mineral shows. (Value range F.)

Another group of calcite crystals with pyrite inclusions from North Aurora, Illinois.

Native sulfur in limestone breccia. Stoneco Quarry, Maybee, Michigan. Euhedral (distinct and well formed) sulfur crystals are found in vugs associated with limestone breccias in the Findlay Arch region, which includes southeastern Michigan. (Value range F.)

A FEW MVT-TYPE MINERALS FROM PARTS OF THE GLOBE OTHER THAN NORTH AMERICA

Celestite (Celestine)

Beautiful blue celestite crystals have entered the mineral market through the Tucson show. These came from a sequence of marine Jurassic rocks that occur in Madagascar, the large island country east of Africa bordering the Indian Ocean. This Jurassic rock sequence also contains many fine fossils, including ammonites similar to those found in southern England, France, and Germany. In Jurassic strata of England occurs what is known as the Portland limestone, after which Portland Cement is named. In the 19th century, this limestone was calcined to make a type of high alumina lime, which, when used in masonry, set-up to become harder than conventional lime mortar. It was similar in composition to the Portland Cement used in the construction

Group of Madagascar celestite crystals.

Celestine "egg" with beautiful blue crystals of this mineral in its interior. Sakoany Mine, Safiaa Region, Mahajanga Province, Madagascar. (Value range E.)

industry of today and from which the name Portland Cement is derived. In this somewhat porous Portland Stone occurred vugs and cavities, which were, at times, filled with celestite (or celestine) crystals, a light blue mineral composed of strontium sulfate. The Madagascar celestite occurrence is similar to this. Another celestite occurrence also occurs in punky limestone of northern Ohio and southeastern Michigan, associated with what is known as the Findlay Arch, as well as on Kelley's Island, a island in Lake Erie. This occurrence is in thick dolomite beds of Silurian age that underlie the flat, glaciated topography of the Findlay Arch, an extension of the Cincinnati Arch of Ohio and central Kentucky.

Moroccan MVT Minerals

The Atlas mountains of North Africa are made up of a thick sequence of Paleozoic sedimentary rocks containing thick beds of limestone. Many of these limestones are renown for their excellent and varied fossils, which have become widely distributed through Moroccan fossil and meteorite dealers. Also found in these limestones are occurrences of

Moroccan bladed barite. Bladed barite, similar to that found in the Washington County, Missouri, barite district. This has come in quantity from the Atlas Mountains of Morocco.

White-bladed barite covered with small vanadinite crystals. Moroccan barite often looks similar to that from Washington County Missouri except it is paler in color. (Value range F.)

minerals sought after and mined for mineral specimens by the rock conscious Moroccans. Calcite and quartz geodes, similar to those of the U.S. Midwest also occur in these limestones as well as do deposits of barite, calcite, and galena, not unlike Ozark occurrences. Zinc minerals are mined near the Moroccan town of Milaban, where they can occur with vanadanite in the form of small, gemmy crystals. Similar vanadanite crystals also occur associated with bladed barite. Occurrences of vanadanite are suggested as being related to the arid climate, which exists (and has existed) since the crystals formed—they always being perched on top of the bladed barite that formed at an earlier time.

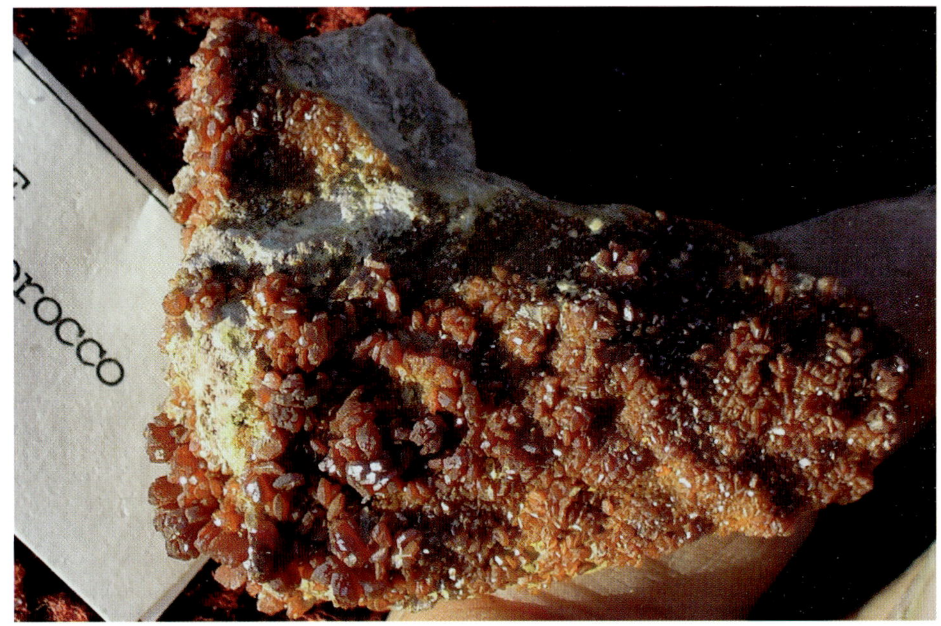

Vanadinite. A beautiful mineral sometimes associated with the widely dispersed Moroccan barite and galena specimens. Vanadinite has not been found associated with Ozark barite or galena. (Value range F.)

Cerussite on galena and sphalerite. Cerussite is a secondary lead mineral associated with a primary one (galena). M'Fis Mine, Mibladin Morocco. (Value range F.)

Vanadinite on barite. M'Fis Mine, Mibladin, Morocco. (Value range F.)

Cerussite on galena. Mibladin, Morocco.

Dolomite with small chalcopyrite crystals similar to the pink dolomite from the Ozarks. Bon Bker, Bouazar, Morocco. (Value range F.)

Single cerussite crystal, Mibladin, Morocco.

Cobaltian-calcite. A large amount of this attractive cobalt-containing-calcite recently (2012) come from Morocco through the Tuscon show and sold at reasonable prices. Similar material occurs in the Ozarks. (Value range G.)

Pine Point NWT Canada Minerals

This is a MVT mineral deposit that occurred in Devonian limestone of northwestern Canada. Essentially a galena, sphalerite, calcite ore body, one of the zinc minerals present besides sphalerite is the polymorph wurtzite. Wurtzite is a zinc, iron sulfide, chemically like sphalerite. Wurtzite in the Pine Point ore body forms botryoidal surfaces with sphalerite, which on a sawed surface produces little rosettes as can be seen here. The mine was closed about 1995.

Botryoidal wurtzite, NWT (Northwest Territories), Canada.

Galena and wurtzite in calcite. Pine Point, NWT, Canada.

More on Supergene (or Secondary) Minerals

When primary sulfide minerals are subjected to weathering and oxidization by being relatively close to the earth's surface, supergene minerals form. Supergene minerals are "secondary" minerals, that is, they formed from what were previously "primary" sulfide minerals, such as sphalerite, galena, and pyrite. As pyrite can be a common mineral occurring in most mineralized areas, when it weathers, it turns into limonite or goethite (a hydrous iron oxide) Associated with supergene minerals can commonly be large amounts of iron oxide. This iron oxide is known as an "iron hat" and is an appropriate name given to the mass of iron oxide associated with them. The "iron hat" is usually composed of limonite, through which, in its lower part, are dispersed the supergene minerals that formed on top of the primary sulfide minerals. The prime Ozark locality for supergene minerals in the southern Missouri Tri-State district was near the town of Granby, especially in what was known as the Beer Cellar Mine. As these were some of the earlier deposits worked in the Tri-State area, few specimens were saved and, as a consequence of this, Granby supergene minerals today are quite pricey and desirable.

Today, specimens similar to those of Granby are available from other parts of the world and these often are as nice as were those from the supergene deposits of the Tri-State district. Many of these specimens come from supergene deposits of Mexico and New Mexico, southwest of the Tri-State district. It's interesting to note that MVT supergene mineralization in the Tri-State district is more varied than other Ozark occurrences, like those of Rush, Arkansas—these minerals appear to take on a southwestern "flavor," much in the same way the modern flora and fauna of the Ozarks of southwestern Missouri starts to take on a southwestern "flavor" with its cactus, yuccas, and other plants along with armadillos and scorpions—critters seemingly more at home in the desert southwest and in Mexico.

Hemimorphite. Ojuela Mine, Mapimi, Durango, Mexico. Some of the mineralization occurring near Granby, SW Missouri, was similar to this. (Value range F.)

White hemimorphite on calcite, part of gossan. Ojuela Mine, Durango, Mexico. (Value range F.)

Hemimorphite on limonite (gossan). Ojuela Mine, Durango, Mexico. (Value range F.)

Hemimorphite. Ojuela Mine, side view—same specimen as above.

Vanadinite on mimetite on limonite gossan. Mapimi, Durango, Mexico. (Value range G.)

Limonite. Part of gossan "iron hat," the oxidized iron derived from weathering of primary sulfide minerals like pyrite. A common occurrence with secondary MVT minerals like those that occurred at Granby, Missouri.

Minetite on limonite (gossan). Mapimi, Durango, Mexico. (Value range G.)

Mimetite on limonite. Mapimi, Durango, Mexico. (Value range F.)

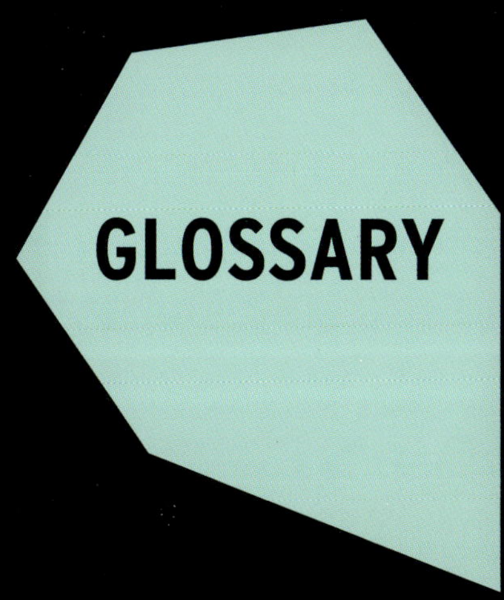

Secondary or supergene minerals

Minerals formed by the alteration of primary sulfide minerals by weathering and oxidization. Supergene minerals are often noted for their rich colors and interesting shapes.

Gossan

When iron minerals like pyrite are subjected to weathering near the earth's surface they produce rust-like minerals like limonite and goethite. Such a concentration of this type of secondary iron minerals is called a gossan, an old German prospectors and mining term. A gossan was (or is) important as such an iron concentration may indicate the presence below it of an ore body of valuable minerals.

Iron Hat

Another term for gossan. Secondary iron minerals may cover a mineable deposit of copper, silver or lead or other valuable elements. It will cover the ore body like a hat—hence the name.

Strontium

An element similar to calcium in its chemical properties but considerably less common that calcium, strontium minerals often occur in limestone and dolomite but for some unknown reason they are almost totally absent in Ozark MVT deposits. Calcium, strontium and barium are in the same group II of the periodic table. It's puzzling that barium is of such frequent occurrence in Ozark MVT mineralization while strontium is essentially absent---its a geochemical enigma.

Vanadium

Vanadium (element number 23) is the most distinctive component of vanadinite which occurs in MVT deposits such as those in Morocco and is extensively illustrated in this book. Its occurrence however may have to do with its association with barite and galena in regions of dry climate. In the Ozarks somewhat soluble vanadium compounds may have been flushed away by this regions extensive groundwater and relative wet environment which did not allow vanadium mineralization to take place.

INDEX